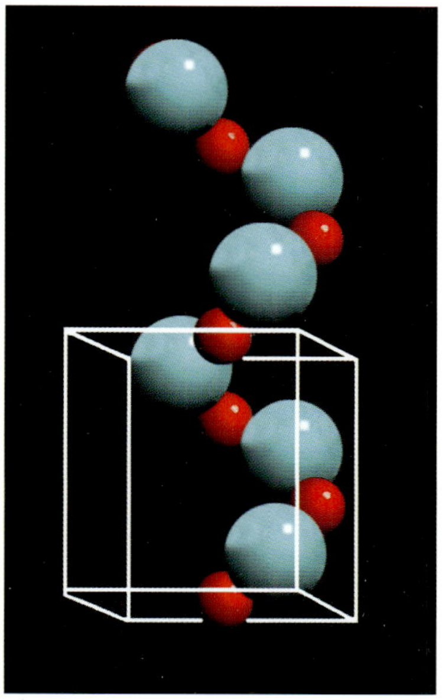

Section I
Internal Structures

Section II
External Forms

Quartz

An Introduction to Crystalline Quartz

Harold L. Dibble

Dibble Trust Fund Ltd.
Angola, NY, USA

Amethyst cluster from Las Vigas, Vera Cruz, Mexico
6.5 cm high Francis Benjamin Collection
Photo by Jeff Scovil ©

Pamela Mills designer

Library of Congress Control Number: 2002115399

ISBN 0-9716025-2-2 (softcover) ISBN 0-9716025-3-0 (hardcover)

Published by Dibble Trust Fund Ltd.
PO Box 67, Angola, NY 14006-0067 USA
dibbletrust@adelphia.net

Printed in Buffalo, New York

PREFACE

People who are interested in the world around them will find that one of the most interesting subjects to study is the mineral species, quartz. There are so many different aspects to the study of quartz and its relatives that everyone can find areas of interest to be delved into deeper. Quartz is such a large component of the earth's crust and so important technologically that we should all have basic knowledge of it.

Quartz is better able than other species to introduce us to the secrets of crystal growth. The generous size and ready availability of quality crystals, their chemical ruggedness and transparency as well as the multitude of variations within the species, make quartz excellent to study. Much of what we know of crystal structure and mineralogy has come through the study of quartz.

Quartz specimens can be intriguing and attractive because of their unusual forms. The great variety of their shapes and colors and relatively low price make them popular items to collect. Crystals are individuals; they were "born" and grew and their shape and appearance were influenced by the environment in which they grew. Despite the enormous number of crystals that exist, no two specimens are exactly alike and there is always something new to be seen or discovered.

The primary purpose of this publication is to open a few doors so that everyone can get a glimpse of the beauty and fascination of quartz. Many studies have been made of the technical details of the different aspects of the species. With our ever-improving technical equipment there come opportunities for further knowledge. A remarkable amount is now known about quartz in all of its varieties and of its related species. There are many different publications on aspects of these studies which are available to those who wish to delve further. Our main interest is in showing the beauty and some of the countless array of shapes and forms and colors of quartz. We also want to point out some of the benefits and important effects that quartz and its relatives have on our everyday lives. It is hoped that this publication will enable everyone to better understand and appreciate quartz and the factors that have led to the specimens that we see today.

To paraphrase Prof. Ichiro Sunagawa, of Tohoka University, Japan: 'crystallography is an exact science on a "dead body" when it is aimed at systematizing the geometry of forms, at analyzing the atomic structure or chemical bonding. However, a crystal can not be formed instantaneously. The conditions of the environment will determine the growth rates, forms, perfections, homogeneities and textures. These conditions may be changed abruptly or even fluctuated. Each crystal has its own growth history and its external and internal forms. Its imperfections and inhomogeneities are a record of that history. Thus we can no longer consider the crystal as a "dead body." Even though small, a stone has its own growth history and individuality. We have been increasing our knowledge and understanding of the relevant phenomena at the atomic level. Our present understandings of the growth and forms of mineral crystals (which are inorganic) are now being used to increase understanding of organic crystals and to build a bridge to the understanding of the living world.'

We want you to enjoy learning about quartz. Many of the subjects covered here are complicated and we have tried to simplify the details. The coverage here is brief but there are many articles and publications which are available to you that cover different aspects of quartz. We recommend that you look up any that may be of interest to you and dig deeper into those phases.

ACKNOWLEDGMENTS

No publication is the work of one person and it is fitting to acknowledge and thank a few persons who contributed so much to making this book as you see it today. The order in which they are mentioned is not to be considered indicative of their contributions to the project.

George Robinson, curator of the A. E. Seaman Mineral Museum in Houghton, MI, USA, made suggestions and emendments that greatly improved the book.

Dr. Gerhard Niedermayr, of the Naturhistorisches Museum in Vienna, Austria, who made several amendments and insisted that all aspects of the subject should receive thoughtful attention.

Prof. Ichiro Sunagawa, of Tohoku University in Tokyo, Japan, for his support of this project and for his lifetime effort to help the world understand and appreciate some of the wonders of the mineral world.

Rudolph Muth whose knowledge and skill with computers and enthusiasm for this publication helped make it possible.

Jeffrey Scovil, whose work is known to almost every mineral lover, and who through his photography helps others to see the beauty in mineral specimens. Rainer Bode, a justly famous mineral photographer. Sally Dibble, my daughter, for her photographic work.

The team of students: Philip, Robert and Eric Niles, who put the manuscript to the ultimate test and also made some excellent suggestions for its improvement.

Much editorial work was done by Richard Bideaux who spent many hours at the task and to whom we are very grateful. Mrs Richard Miller who also made important corrections.

Ernst Both and the Buffalo Museum of Science for the use of one of their microscopes.

Pamela Mills, the publication and design consultant, who insisted on professional quality work and whose expertise greatly improved the book.

Many others too numerous to mention, for their support and helpful suggestions.

Harold L. Dibble

This publication is dedicated to

Doris I. Dibble

who died in April 2001

She served as Secretary and Treasurer of

the Dibble Trust Fund Ltd.

Her interest in the project and her

wise judgment in many matters are missed.

The Dibble Trust Fund Ltd.

CONTENTS

Preface 3

Internal Structures

Introduction to Silica 7

Polymorphs of Silica 9

- Quartz 9
- Tridymite 9
- Cristobalite 9
- Stishovite 9
- Keatite 10
- Moganite 10
- Melanophlogite 10
- Lechatelierite 10

Studies of Crystal Growth 11-13

Left and Right Handedness 15-17

Nucleation and Growth of Crystals 19

Nucleation 19

Growth 20

Conventional Crystal 21

Growth Mechanisms 23-24

Twinning 25

Dauphiné Twinned Quartz 25-27

Brazil Twinned Quartz 29-31

Piezo-electric Effect and the Commercial Uses of Quartz 31-32

Color 35

- Smoky Quartz 35
- Rose Quartz 35-37
- Citrine, Amethyst and Ametrine 37
- Blue Quartz 37
- Green Quartz 37

Crystal Shapes 39

External Forms

Types and Forms of Crystals 41

Tapering and Expanding Crystals 43

Tessin Habit 43

Artichoke or Rocket Habit 43

Muzo and Cipo Habits 45

Faden Quartz 47-48

Crystals with "Children" 49

- Cathedral Quartz 49
- Pineapple Quartz 49
- Pinetree Quartz 50

Altered Terminations 51

Scepter Quartz 51

Window Quartz 54

Inverse Scepter Quartz 56

Skeleton or Hopper Quartz 56

Japan Law Habit 60

Interrupted Crystals 62

Quartz "Diamonds" 64-65

Gwindel Quartz 68

Inclusions 72

Phantom Crystals 74

Péle de Jacaré or Crocodile 75

Clusters of Tabular Crystals 75

Haystack or Jackstraw Quartz 76

Globular Quartz 77

Plates of Tiered Crystals 80

"Odd Balls" 80

Parallelogram 80

Iris or Rainbow Quartz 81

Hemimorphic Type Crystals 83

Bilateral Parallel Quartz Druses 83

Gemstones and Novelty Items 88

Crystal Face Surfaces 92

Properties of Quartz Table 93

Contributors 94-95

Bibliography 96-97

Index 98-99

Geological Dictionary 100

Amethyst cluster from Amatitlan, Querétaro,Mexico 26.3 cm high
C. G. Clark Collection Photo by Jeff Scovil ©

Introduction to Silica

Back at the beginning of time when the Earth, solar system and universes were being put together, their creator had some fun. Instead of blocks or cubes, a tetrahedron was created with four equal triangular sides. It also had a little glue on the corners so that other units could be attached in various ways at any two corners. The purpose of this publication is to give you some idea of the many different ways that these can be and have been put together and to give you enjoyment from the appreciation of the colors, shapes, varieties and habits that have resulted.

To understand the basic structure of silica, let us consider three spheres placed together to form a triangle.

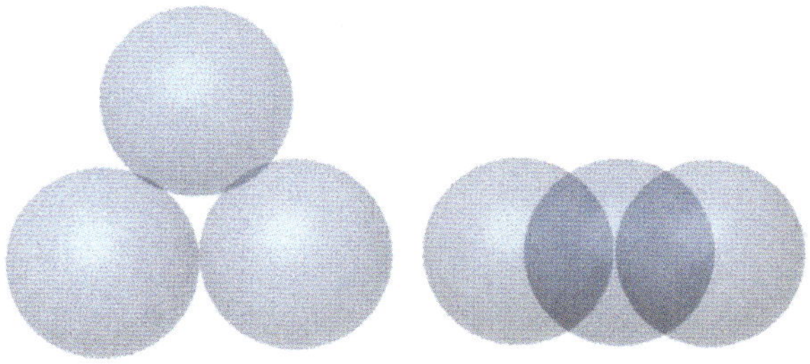

Top View Side View
Fig. 1

The large spheres shown represent oxygen. There exists a small space between the spheres. Into that small space introduce a small sphere of a different element, one with a powerful attractive force - silicon.

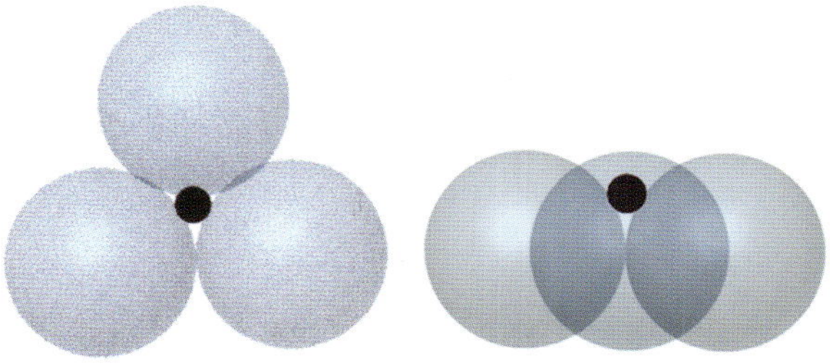

Top View Side View
Fig. 2

On top of the center of this triangular arrangement place another atom of oxygen. It will fit nicely onto the depression between the three large atoms and be held firmly in place by the silicon's attraction.

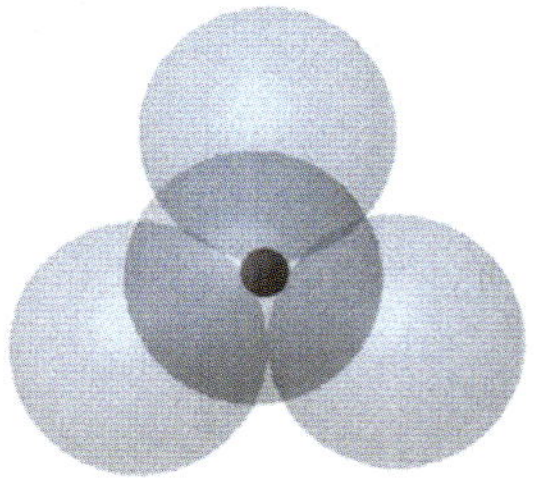

Top View
Fig. 3

Side View
Fig. 4

This essentially is what the silica tetrahedron looks like. The attraction between the silicon and the oxygen atoms is strong and the distance between them so small that the pull they put on each other is intensified. This intense attraction is what makes quartz so durable and thus important to the world. The simplicity and strength of the silica tetrahedron are reasons for many of the attributes of quartz and its polymorphs.

Another factor that makes quartz so important is the enormous quantities that exist. It has been estimated that quartz makes up approximately 12 percent of the earth's crust. When the weight of the silicon and oxygen in all the other silicate species are added to this total (that is, species of minerals which consist of silicon and oxygen and a few other elements) the total goes up to almost 84 percent of the earth's crust.

The simplicity and strength are particularly responsible for the purity, clarity and beautiful crystallization of quartz. These factors in turn have led to the choice of quartz as the species to be most often studied. Much of what we know today about all minerals has come about through the studies that have been made on quartz crystals, particularly the clear variety known as rock crystal.

Beta quartz from 2nd Sovietsky Mine, Dal'negorsk, Primorsky Kray, Russia 5.2 cm high
Jeff Scovil Collection Photo by Jeff Scovil ©

Polymorphs of Silica

There are many different ways that the tetrahedrons of silica may be arranged. What arrangement is actually used will depend on the temperature and pressure at the time of formation. Fig. 5 is a chart showing different species that are created of the silica tetrahedrons and it shows the temperatures and pressures at which they take place. The tetrahedrons are all the same, only the arrangement is different. The tetrahedrons are connected at all corners with each oxygen atom being part of two tetrahedrons. The result of this is that there are only two times as many oxygen atoms as silicon atoms in quartz and its polymorphs and the chemical formulas are all written as SiO_2.

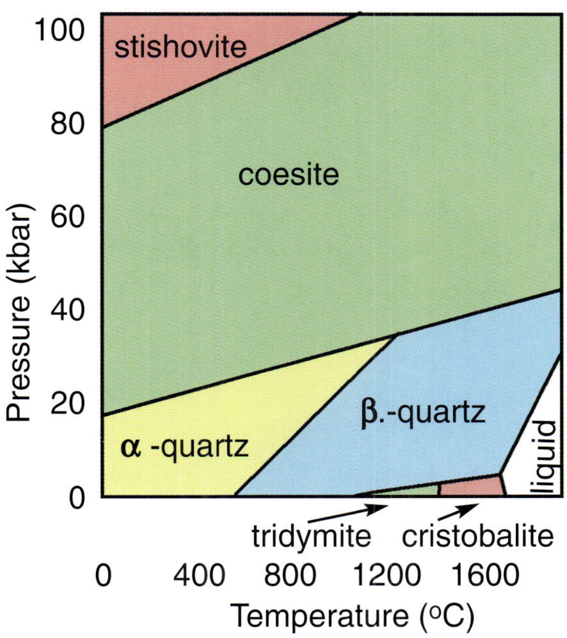

Fig. 5

The structure of each of these forms of silica is stable within the indicated temperature and pressure ranges. If the conditions change and move outside the original range, the compound becomes unstable (metastable is the word used). If the change in the connections between the atoms is minor as is the case between α ("low") and β ("high") quartz, the change can take place almost instantly. If, however, the bonding requires a large change, such an alteration could take years to occur.

Quartz α and β

In the β range the tetrahedrons form a ring of six units in a flat plane. In the α range, there is not as much energy in the forces which repel similar atoms and the attractive forces pull the units closer together. The tetrahedrons are twisted through 16.3^{o} and the rings are "wavy". In the other direction (perpendicular to the plane of the rings) the atoms attach themselves together in the form of a helix or corkscrew.

Tridymite

Is said to exist in five separate phases not just two like quartz.

Cristobalite

Like quartz also has two structural arrangements α and β. Both tridymite and cristobalite have been found in volcanic rocks, meteorites and lunar rocks.

Stishovite

The Max Planck Society reports that "The synthesis of stishovite and its subsequent discovery in naturally shocked rocks in Meteor Crater, Arizona, USA, and the Ries Crater, Germany, has revolutionized the study of shock metamorphism and impact cratering by providing an index mineral in addition to coesite, that can be used as proof of shock metamorphism."

Notice that it was only after laboratory work creating stishovite and studying it scientists were able to identify it in the field. Note too that the crystals did not take thousands of years to nucleate and start growing, but when the temperature and pressure conditions needed were suitable, the crystals were created in the moment of impact.

Keatite

One arrangement of silica which is not shown on this chart is Keatite. This polymorph has been produced in laboratories by subjecting silica to high temperatures and pressures. It is not listed as a species of mineral because it has never been found in nature.

Moganite

This is one of the latest members of the family. Drs. Peter Heaney and Jeffrey Post have been leaders in the research on this "naturally abundant but little known member of the silica system." They report "It is now recognized as a common intergrowth within unaltered microcrystalline varieties, such as chert and chalcedony." It is estimated that 10% of agate is moganite. It is not recognized as a separate species.

Melanophlogite

This polymorph has an unusual open structure and all known specimens of this mineral have organic compounds in some of those openings. For that reason, melanophlogite has been approved as a species defined as "SiO_2 with organic compounds." It is not known if the organic compounds have had a role to play in the formation of its structure.

Lechatelierite

Is included in some lists as a polymorph of quartz and although it is almost pure SiO_2 it has no internal structure. It is, therefore, a glass and is not recognized as a mineral species but is classed as a mineraloid.

Beta quartz cluster 14.5 cm wide 2nd Sovietsky Mine, Dal'negorsk, Primorsky Kray, Russia
H. Dibble Collection Photo by Sally Dibble ©

Silicon dioxide polymorphs are an interesting lot and much could be and has been written about each of them. Since this publication concentrates on quartz, you are referred to other sources of information on the other family members. Web sites, in particular, offer enormous amounts of information. Many new sites covering minerals are being opened thus offering a wealth of knowledge. Some have interactive applets that enable better understanding of the three dimensional nature of the different structures. Because of the rapidly changing nature of the medium we have not suggested a list of excellent sites. It is suggested that you use a good search engine to locate the information that you want and then try several sites. You should expect to be amazed and pleased at what you will find.

Studies of Crystal Growth

Why crystals look as they do is a subject to which many persons have devoted years of their lives. As we learn more, and as advanced technology permits us to learn more, new areas of study open up. It is an ongoing process.

Because the number of persons who have contributed to our understanding of quartz is so large and the nature of their contributions so varied that their story cannot be told in a few paragraphs. No compilation can ever tell the complete story because it involves many individuals over many years and in many countries looking at different aspects of the mineral species. The story, however, is interesting.

It seems humorous to us now that the ancient Greeks and Romans and persons who lived in the Middle Ages believed that clear rock crystal quartz was ice that had frozen so hard that it could not melt. This error was easy to make because after all, the two look somewhat similar, and quartz is cool to the touch. That was enough information for them. The first scientific description of quartz came in a publication by Nicolaus Steno, which was published in Copenhagen in the year 1669. He pointed out that there was a constancy in the angles between the corresponding faces of quartz crystals. Other scientists soon followed his lead in studying these angles, including Rome de l'Isle (1736-1790) and René Haüy (1743-1822). They added to and refined his rules. Steno also added another important first, for he stated in his publication that crystals grew by the addition of particles to their surface.

In the years that followed, understanding the exact arrangement of the atoms in the molecule and the arrangement of the molecules to each other remained one of the big problems. One way scientists tried

Cluster from Conselheiro Pena, Minas Gerais Brazil 8.7 cm high
Francis Benjamin Collection Photo by Jeff Scovil ©

Cluster from Petersen Mountain, Washoe Co., NV, USA 10 cm high
Hallelujah Mining Collection Photo by Jeff Scovil ©

Open gwindel from North face Les Droits, Argentière, Chamonix, Haute Savoie, France
Eric Asselborn Collection Photo by Jeff Scovil ©

to understand this was by measuring and recording the angular positions of all the crystal faces. After they had measured the most common angles they had not solved the question of arrangement, so they continued the measuring on smaller and smaller faces including some that they were not even sure of and disagreed about. They still had no answer. The big breakthrough came about a century ago with the work of the Curies, Marie and her husband Pierre, who with Henri Becquerel discovered radioactivity and thus proved that the atom is not the smallest indivisible unit. After the discovery of X-rays, scientists used them on crystals to produce diffraction patterns. This led to the ability to determine the position of atoms inside molecules.

Many new types of scientific equipment have been invented and developed over the last century. Some of the equipment now in use includes mass spectrometers of various types, nuclear magnetic resonators, ion chromatographs, scanning electron microscopes and electron microprobes. These devices and others have enabled and enriched our understanding of minerals to a point that could only be dreamt about in previous times.

Computers too are increasing our knowledge today in a multitude of different ways. There are many different computer programs covering many aspects of mineralogy and more are being written all the time. For example, reports by Müller-Krumbhaar et al and Gilmer and Jackson in 1977 based on computer simulation programs were able to add to our understanding of crystal growth under different environmental conditions. In a short time after that, computers were able to simulate growth patterns that previously would have taken years to have been produced in a laboratory.

Amethyst from Bellingham, MA, USA
6 cm high
Bob Borofsky Collection Photo by Jeff Scovil ©

Cluster from Aiguille Pierre a Joseph, Talefre
Mer de Glace, Haute Savoie, France 6.6 cm wide
Francis Benjamin Collection Photo by Jeff Scovil ©

Citrine cluster from Minas Gerais, Brazil 9 cm high
Martin Zinn Collection Photo by Jeff Scovil ©

Left and Right Handedness

At this point, another concept needs to be introduced, unit cells. A unit cell of quartz consists of three molecules of silica, that is, three atoms of silicon and six atoms of oxygen. There is one atom of silicon on each of the six faces of the cell; but as the atoms are only half into the cell and half into the adjoining cell they only count as three. Two sides of the cell have a silicon atom in the middle of its face. If the cell is placed in such a position that one of those sides is nearest the observer, then the face furthest away will also have the silicon atom in the middle of its face. Of the remaining sides, two opposing faces will have the silicon atom one third of the way back and the last two faces will have the silicon atom two thirds of the way back. The six atoms of oxygen in the cell fit between the silicon atoms and unite them. The size of the unit cell is 4.91 x 4.91 x 5.40 Å (an ångstrom unit is equivalent to one ten millionth of a centimeter).

The silicon atoms can connect themselves to the oxygen atoms in two different ways. Once connected, the atoms will form a helix (as was mentioned on page 9).That helix will be either left or right handed in the same way that left and right hands are the mirror images of each other.

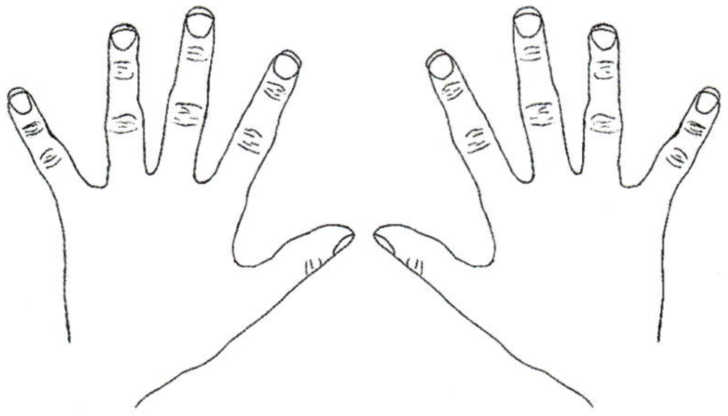

Left hand　　Right hand

This is a three-dimensional arrangement that is difficult to represent on a two-dimensional surface like paper. It can be explained as a left or right hand corkscrew. Another way to visualize the arrangement is to consider that you are standing in the center of the bottom of a "spiral" stairway and look straight up; a left hand helix would rise clockwise and a right hand helix would rise counterclockwise.

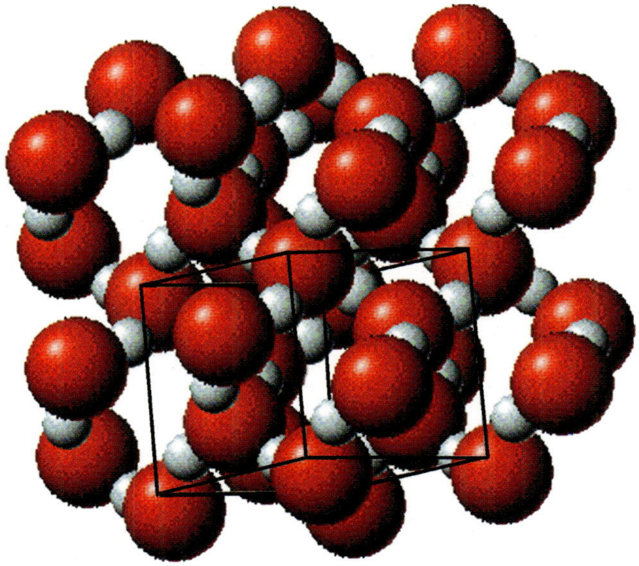

Si
O

Left-handed quartz
Fig. 9

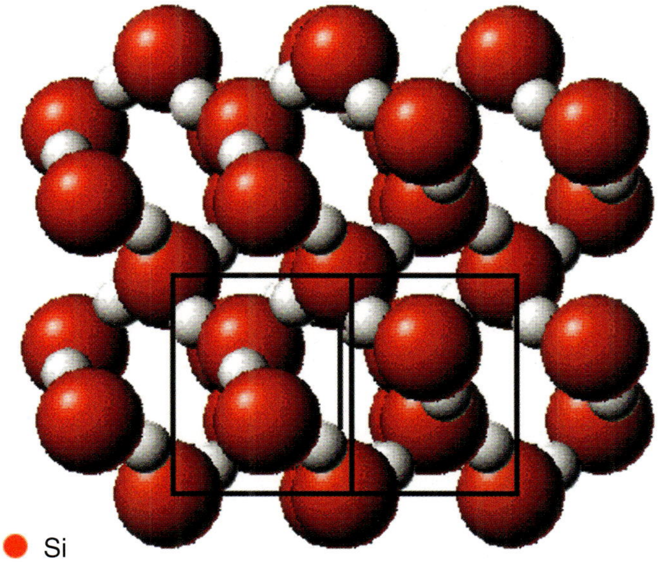

Si
O

Right-handed quartz
Fig. 10

To further help clarify this explanation are these representations of left and right quartz with unit cells shown.These illustrations are based on the applets on the CD that accompanies this book. Manipulating these applets helps visualize the three dimensional arrangement of the atoms even better than these pictures.

Note: The ratio of the size of silicon and oxygen is not correct as shown. It has been changed in order to best illustrate the relative positions between of the atoms. If correctly shown the silicon atoms would be difficult to see. The relative sizes of the atoms are correctly shown on page 7.

"Cathedral" type cluster from Brazil 16.9 cm high
Francis Benjamin Collection Photo by Jeff Scovil ©

Here are two representations which show only those atoms and their connections which are involved in the helical groupings.

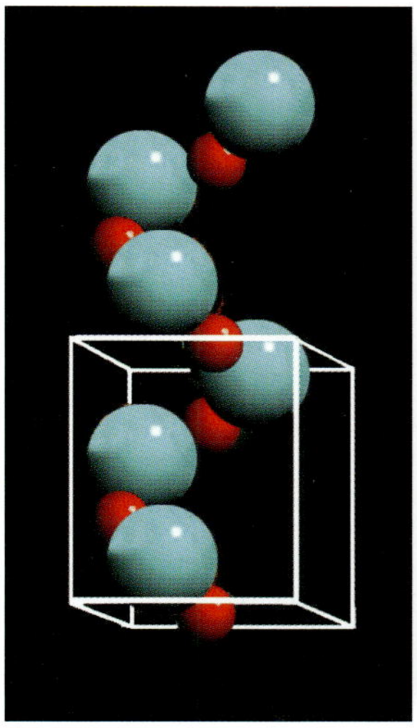

Fig. 7
Left-handed quartz

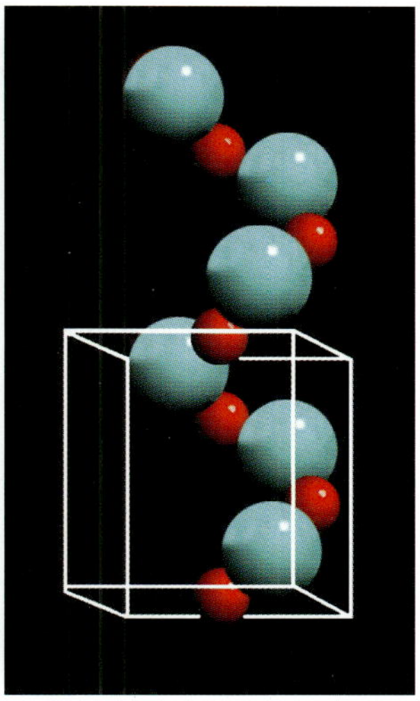

Fig. 8
Right-handed quartz

In the figures 7 and 8, silicon is represented as red and oxygen as blue.

From the Wah Wah Mtns., UT, USA 9.3 cm high, Martin Zinn Collection Photo by Jeff Scovil ©

Quartz with hematite inclusions from China 6 cm high
Wydar Collection Photo by Jeff Scovil ©

Quartz from 8-Meteor Pocket, Silverton, CO, USA 20.5 cm high
Benjy Kuehling Collection Photo by Jeff Scovil ©

Smoky quartz cluster from Minas Gerais, Brazil 17.5 cm high
H. Dibble Collection Photo by Jeff Scovil ©

Nucleation and Growth of Crystals

Prof. Ichiro Sunagawa of Tohoka University classifies growth of crystals into either solid-solid or liquid-solid crystallization. He then subdivides the latter into melt growth, solution growth and vapor growth. He states, "The most important and universally encountered growth of crystals is solution growth." So in considering nucleation and growth here, we will assume them to take place in a solution.

Several different factors affect the growth rate of quartz crystals in a solution. Among these are the concentration of silica, the temperature, the pressure, and the speed and direction of the flow of the solution (if any). Impurities also affect the growth because the presence of other chemicals may act as deterrents or catalysts to growth. It is the changes of the temperature and pressure of the solution, however, that are most important. For example, if the solution has carried a large number of silica units and the temperature and/or pressure drops, the energy level drops and the carrying capacity of the solvent drops. The solution becomes "super saturated" and the solute (the particles dissolved in the solution) start to unite and nucleation will begin.

Nucleation

The word nucleation as it is used here is the concept that individual molecules or tetrahedrons of a substance (in this case quartz) will start linking themselves into larger units, that is, into unit cells and clusters or attach themselves to any suitable location (such as a similar unit or an already growing crystal).

New quartz crystals can start in one of three ways:

1. A coming together of floating particles, a sort of coagulation.
2. Attachment to an anchored or fixed particle of quartz.
3. An epitaxial alignment of the new quartz particle on a particle or crystal of a different species.

Perhaps the word epitaxy needs to be explained at this point. It means a growth of one mineral species on the surface of another species, so that the unit cell of the new mineral (which is being moved by electromagnetic forces) will be aligned with the cells of the original mineral. [*It is a word that has come into frequent use recently particularly by those working on layered computer chips. They need to achieve lateral growth of a layer of one mineral species which can be active on the surface of another species which is inactive.*]

The types of quartz crystals affected by layers of other species are the ones described in this book as *altered terminations*, *interrupted growth* and perhaps *faden crystals*. When an existing quartz crystal has been covered by only a thin layer of another species (usually calcite or a clay mineral), the forces of the old underlying quartz crystal usually operate through the thin layer and determine the alignment of the new quartz particles. In addition the axis of the new crystal will be parallel to that of the older crystal.

In a different situation, when there are calcite crystals present in the area, even without underlying quartz, the calcite crystals can still serve as bases and determine the alignment for new quartz nucleation. Dr. Ivan Bonev, of the Geological Institute of the Bulgarian Academy of Sciences, points out that while the structural elements and the general structural patterns of quartz and calcite are quite different, the cross section of both, taken perpendicular to the ***c*** axis, are hexagonal. In addition, the size difference is 1% or less, so quartz particles can easily attach to existing calcite crystals and align with their crystal structure.

Growth

One explanation of the different rates of growth in different directions on a crystal was given in 1955 in an article by P. Hartman and W. G. Perdok and they used a diagram based on a cubic crystal.

The smooth surfaces called *faces* (f) that are shown in blue can attract particles along only one axis. The areas that attract along two axes [here shown in green] are called slopes (s). Those areas that attract particles along three axes [here shown in red] are called kinks (k).

Each molecule of each species (quartz for example) has a particular size and shape and each molecule is attracted to other similar molecules. Three molecules of quartz unite in a unit cell, one side of which is longer than the other two. The energy of attraction will vary in strength among the different axes of the unit cell. When similar unit cells are near each other they tend to align themselves so that the axes of the strong attraction forces line up with each other and the weaker attractions axes line up also. When the actual joining or bonding of the cells takes place, energy is given up to the surrounding solution and this helps lock them in place.

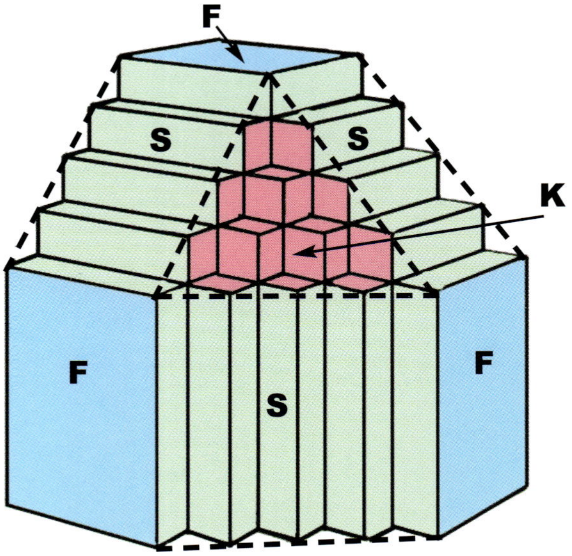

Because some faces of the unit cells are longer than others and the strength of the bonding energy differs, growth of the crystal will differ in different directions. The *Periodic Bond Chain Mechanism* diagram is an attempt to portray the action that takes place.

See the periodic bond chain mechanism applet on the CD to view a video of the action that takes place.

In the attached applet, notice that the shape of the crystal changes as it grows. It is ironic that the fastest growing faces (the kink surfaces) disappear instead of growing larger. They grow themselves out of existence.

Periodic Bond Chain Mechanism Diagram

A simplified diagram showing how some crystals grow

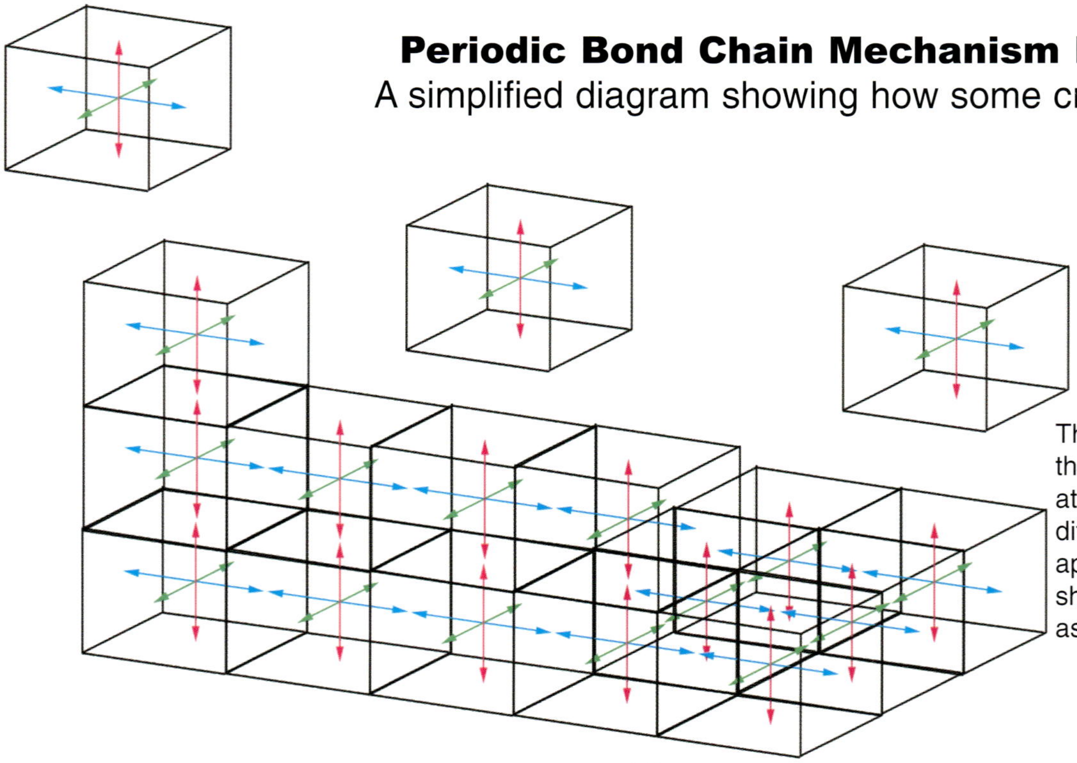

The arrows represent the strength of the attraction along the different axes. The approaching cell should be considered as in solution.

Conventional Crystal

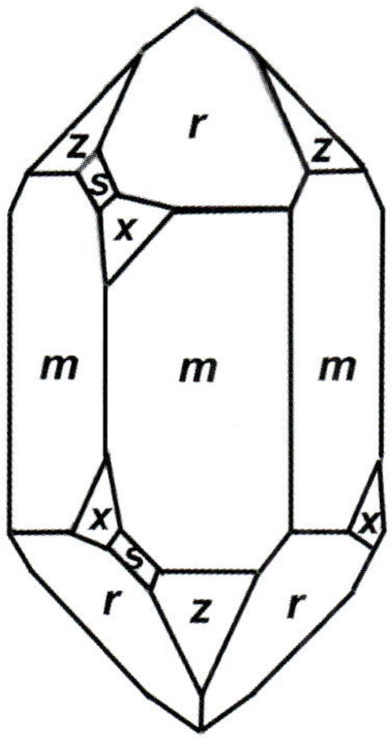

Fig. 11

Years ago in order to make discussion about crystals easier, crystallographers adopted certain conventions. The illustration (Fig. 11) shows the letters used for some of the faces. The letter ***r**** is used for the largest of the terminal faces, the positive rhombohedron.

If three of the terminal faces out of six are larger than the other three, then all of these three are considered as ***r*** faces. If all six faces of a termination are of equal size, then they are examined for a difference in their appearance. If there is a difference in their luster, the shinier ones are designated as the ***r*** faces.

The letter ***z*** is used for the other terminal faces, which are called the negative rhombohedrons.

The letter ***m*** is used for the prism faces, the faces that are parallel with the **c** axis (the long axis).

The letter ***s*** is used for a trigonal dipyramid face with parallel sides.

The letter ***x*** is used for a trigonal trapezohedral face with unequal size edges.

The crystal shown above is a left-handed crystal.

There is not universal agreement on how to classify quartz. It has a trigonal symmetry which in the United States is considered a sub category under hexagonal. In Europe trigonal symmetry is considered a separate category. A crystal like this would be called pseudo-hexagonal.

*[While we make use of conventional symbols, we break with convention and use **bold** type for those symbols to enable faster and easier reading]

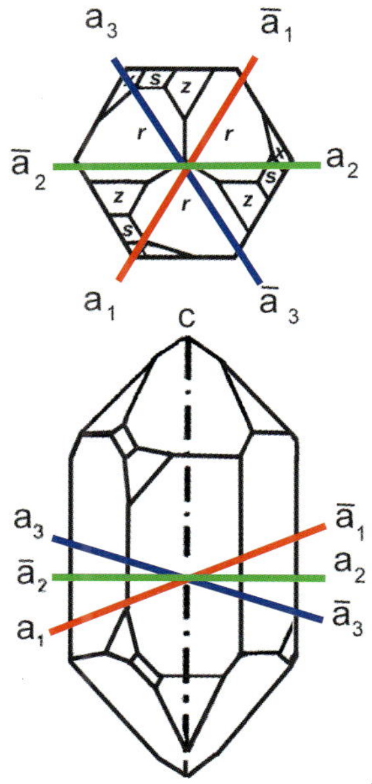

Fig. 12

The above are orthogonal views of the conventional crystal with a view from the top and from the side to better represent the axes of the prism. The bar over the letters indicates the negative end of an axis.

Specimen from Meteor Pocket, Ouray, CO, USA
30.5 cm high
Benjy Kuehling Collection Photo by Jeff Scovil ©

Gwindel from Riante Vallee, Les Courte Argentière, Chamonix, Haute Savoie, France 7 cm wide
Eric Asselborn Collection Photo by Jeff Scovil ©

Growth Mechanisms

Because the attraction of new particles to a growing crystal will be different in different directions, some surfaces have more capacity to attach new units than do other surfaces. They will therefore grow faster. The surfaces with the fewest atoms per unit of surface will fill in fastest. The surfaces with the most atoms will be the most stable, having the most bonding requirements satisfied, and are generally the slowest to grow.

Several different theories and factors are now used to explain exactly what happens. The principal mechanism that the growing crystal will actually utilize will be dependent on the amount of energy available. This, in turn, is dependent on several factors including temperature, saturation, the distance of the particles from a growth surface and the roughness of that surface.

Two principal methods of growth are known as the VKS and BCF growth mechanisms. They are named for the scientists who made major contributions to the study of this subject: M. Volmer, W. Kossel and I.N. Stransky in the first instance, and K. Burton, N. Cabrera and C.F. Frank in the second instance.

On smooth planar surfaces, energy to attach new particles is available from only one direction. Once an attachment is made to that surface, at whatever level of size, a raised hillock is formed and energy from two directions, not just one is available to attract additional growth units, and this will speed growth. Then, if there is growth of the hillock in a second direction along the smooth surface, a third attracting face will be formed and this will result in even faster growth. This is the reasoning behind the VKS mechanism.

See CD at back of book and view applet showing the VKS mechanism.

The BCF growth mechanism is known as the spiral growth mechanism. It is triggered by the misaligned attachment of a cluster of unit cells so that they in effect form a ramp on an otherwise smooth surface. The side of the ramp forms the second attracting face and as soon as a unit cell attaches itself, a third attracting face is formed.

The growth continues along the lowering side of the ramp. At some point, new growth units "turn the corner" and start back up the ramp. At the top of the ramp the new growth again turns the corner and a spiral is started. When the greatest attraction is perpendicular to the surface of the original crystal, needlelike growth results. When the greatest attraction is parallel to the surface of the original crystal, sheetlike spiral growth develops.

To view BCF mechanism, see applet on the CD.

One of the interesting aspects of this spiral growth is that the shape of the growth varies according to environmental conditions such as supersaturation of the solution and/or temperature. In recent years, scientists have been able to devise computer programs that can show the growth of crystals under different conditions. They have studied the growth with varying saturations of quartz solutions under different temperature and pressure conditions. They found that in some cases when the temperature is fairly high, the growth will form a true spiral and in other cases when the temperature is lower, it will have a polygonal pattern. This growth is often referred to as two dimensional nucleation growth, but it does have a thickness usually of at least 4.9Å units between steps.

Much crystal growth in a solution originates at the edges of faces for more than one reason. One reason is that there are always open kink sites along the edges where the faces meet, i.e. where a termination face meets another termination face, between termination faces and prism faces, and between prism faces. These sites are open because unit cells of quartz cannot align themselves perfectly where edges meet.

A second reason is that there are more floating particles available for growth there.

The reasons for this were explained in a paper published by W. F. Berg in 1938. He indicated that the solution around a growing crystal is not uniform in saturation. The concentration of particles in the solution available for growth is greater at the edges of a flat face than it is in its center. In a solution in contact with a flat face of a growing crystal, the particles being added to the crystal come from the area closest to its surface. When particles leave the solution and attach themselves to the crystal the supersaturation of the solution is decreased in that area.

Fig. 13 Diagram of Berg Effect

The dark square in center represents a crystal. The lighter blue areas surrounding the square represent the growth solution surrounding it. The lighter colors of the solution indicate the parts of the solution that have lost silica into the growing crystal. You will notice that the darker sections of the solution are closer to the crystal at its edges than at the centers.

One example that Berg used represented a crystal with a 90° angle between the faces. In such a situation the points on the edges between the faces would have a 50% greater number of particles available for attachment than a spot in the middle of a surface. For quartz crystals with 120° internal angles the number of such particles would be one-third greater.

A second effect is caused by the attraction that exists between the floating particles of silica and the surface of the crystal. Particles which are already on a surface, even if they are not bonded to it, are attracted by it and will not move along it unless acted on by a strong force. This attraction is referred to in hydraulics as "friction".

The third factor which alters the rate of growth is the circulation currents in the solution which affects the dispersion of the particles in it. Strong currents tend to equalize the saturation in the solution. They will reduce the width of the areas near the crystal that have been depleted of solute by the incorporation of some of the silica into the crystal.

If the growth is slow, crystals will form more perfectly. The arriving particles seem to push impurities out of the way. Most of the additions are of unit cell size, and growth is orderly. If, on the other hand, rapid drops in the temperature or pressure of the solution create a situation of great supersaturation, the growth pattern may be badly broken up. Several different things may happen. It is remindful of the children's game "Going to Jerusalem" or "Musical Chairs". That is a game in which children move around in a circle to the beat of a march tune and there is one less chair in the center of their circle than the number of children. When the music stops all try to find a chair to sit in and there is a wild scramble. One result of the "mad scramble" of unit cells for incorporation into the crystal is the incorporation of material which is not a perfect fit; it is like fitting a large child in a small chair. For example: tetrahedrons with aluminum and not silicon in their center are larger and don't really fit properly. They may become part of the growing crystal anyhow. Then too, there are impurities which become imbedded into the crystal because the quartz simply grows around them.

In this discussion of growth, it should be noted that if the conditions of the solution change and more energy is added by increasing temperature or pressure, the process of growth may be reversed. Unit cells will free themselves from the crystal and leave it with a dull matte finish or even pitted.

Twinning

As we saw, depending on which silicon atoms bond with which oxygen atoms, 'right' or 'left' unit cell arrangements may be formed. As crystals grow, right-hand units usually unite with right-hand units and left with left. That way, because of the position of the atoms of silicon, the units fit together perfectly. If the growth of the crystals is slow and orderly, crystals will form that are composed entirely of 'right' or 'left' hand cells. As these crystals grow there are certain locations that grow the slowest and on some crystals small faces develop at those locations. These are the faces that have been designated as ***s*** and ***x***. Figure 14 shows idealized forms of right and left hand doubly terminated crystals with the faces designated as ***s*** and ***x*** well developed and which will indicate the "handedness" of the crystal.

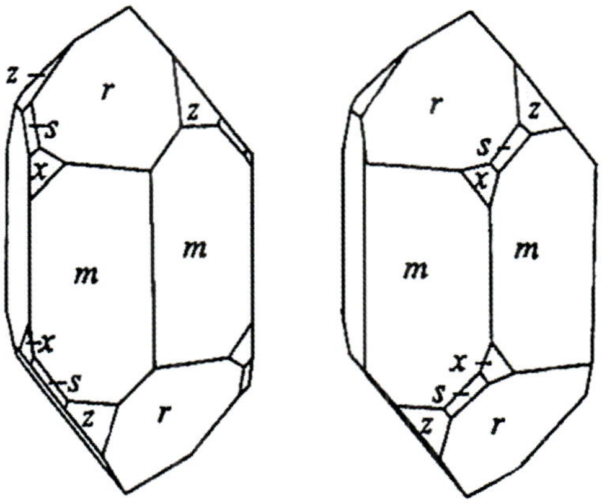

Left-handed Right-handed

Fig. 14

If the growth rate of the crystal is rapid, there are several different things that may happen. In some cases groups larger than unit cells, such as clusters or crystallites, may land on a growing crystal and create a new center for growth.

Dauphiné Twinned Quartz

If the new group is of the same hand and intergrowths develop, the crystal would be called a Dauphiné twin and may look like the crystal represented by Figure 15.

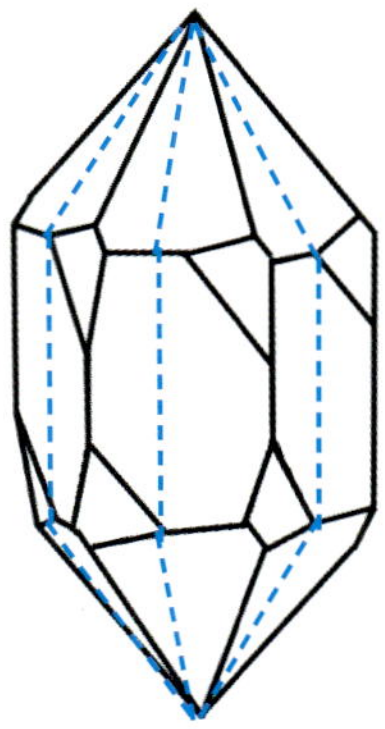

Fig. 15

It can best be thought of as an intermingling of two crystals with their vertical axes as either the same or parallel. One of the crystals has been rotated half way around (180°) so that one of its ***z*** faces in the top termination now is on the same plane as the ***r*** face of the other crystal.

The irregular areas of high and dull luster indicate Dauphiné twinning in this Tessin habit crystal

Specimen from Dara Ismael Khan Dist. Waziristan, Pakistan 8.9 cm high
In a private collection Photo by Jeff Scovil ©

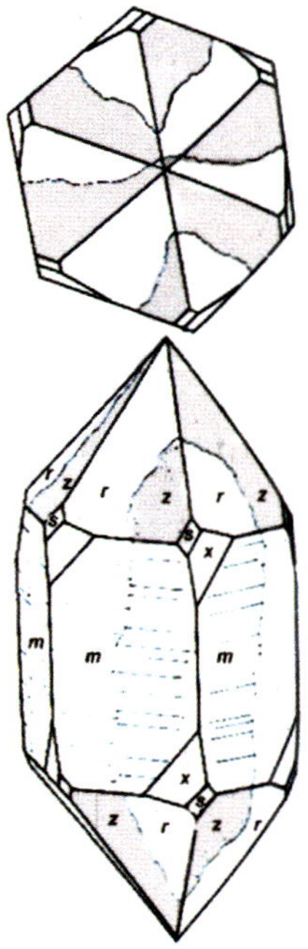

Fig. 16

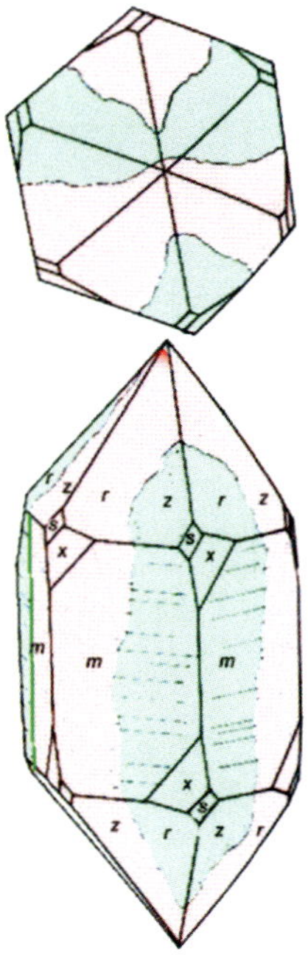

Fig. 17

One of the ways that is used to determine if twinning is present in a quartz crystal is to etch it with the proper solution. The ***z*** faces will etch more deeply than the ***r*** faces. A double "terminated" left dauphiné crystal might, therefore, look like Fig. 16. If this crystal could be colored with one left handed twin colored red and the other green, they might look like Fig. 17.

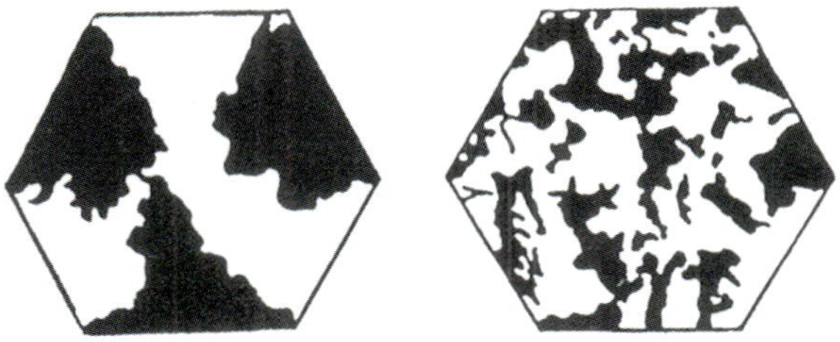

Fig. 18

If the growth of Dauphiné twin crystals is slow, the intermeshed crystals tend to keep separate and cross section views of these crystals will show that three alternate sectors are dominated by one crystal and the other three are dominated by the other "hand", as is illustrated by the left hand

Fig. 19

sketch by Rudolf Rykart in Fig. 18. When the growth of the crystal is more rapid, the sections of each become smaller as is shown in the sketch on the right of Fig.18. Rykart calls this type *macromosaic*, or "Friedlander" quartz. Fig.19 shows photos of this type with the different sections being noticeable because of the difference in the luster on the faces.

Citrine crystal from Brazil
field of view 9 cm high
Surface shows Dauphiné mosaic twinning

Cluster of quartz crystals with muscovite from Minas Gerais, Brazil 24.8 cm high
Wayne Thompson Collection Photo by Jeff Scovil ©

Tessin shape quartz crystals with hematite inclusions on sphalerite 8.8 cm high from 2nd Sovietsky Mine, Dal'negorsk, Primorsky Kray, Russia
Terry Huizing Collection Photo by Jeff Scovil ©

Smoky quartz crystals with the amazonite variety of microcline 11 cm high from Tree Root Pocket, Two Point Claim, Teller Co., CO, USA
Dave Bunk Collection Photo by Jeff Scovil ©

Unusual specimen with two Japan law growths but the largest part has hexagonal cross section from Andilamena, Tamatave, Madagascar 5.3 cm high
Tropic Stone Collection Photo by Jeff Scovil ©

Brazil Twinned Quartz

If the new growth attaching itself to the original crystal is of the opposite hand, the resulting crystal might look like Fig. 20. This type of twinning is known as Brazil twinning. Brazil twins may show even greater complexity than Dauphiné crystals in the way the two sections integrate.

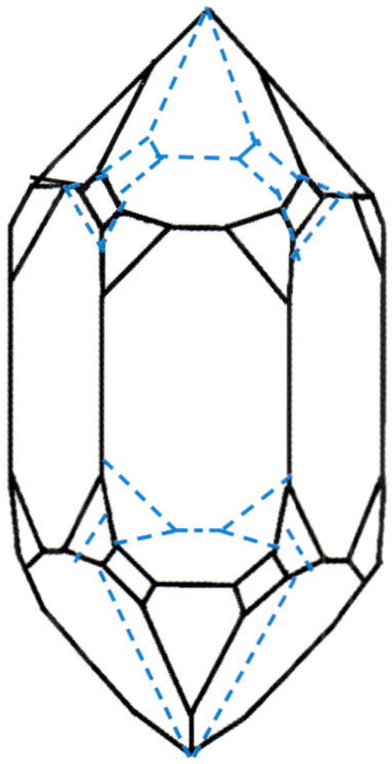

Fig. 20

Prof. I. Sunagawa illustrates, in the drawing Fig. 21, one example of an unusual effect of twinning which is found in some amethyst crystals - the Brewster fringes. Fig. 22 is a photograph of a cross section of an amethyst crystal with these "fringes".

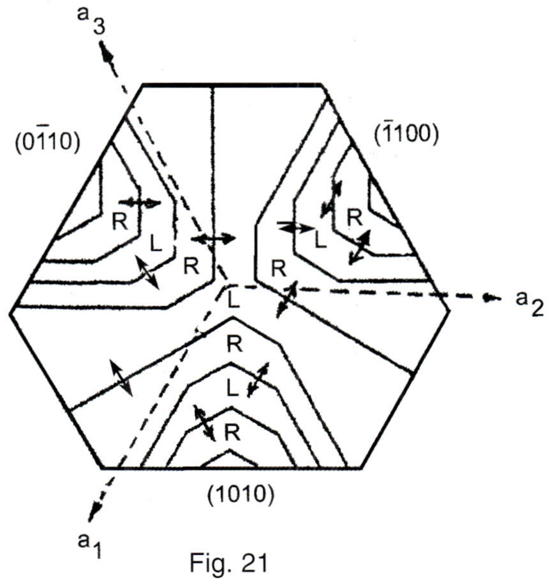

Fig. 21

Schematic illustration of Brewster fringes in a slice of amethyst under crossed polarizers

Fig. 22

This photograph is from a book on amethyst by Werner Lieber. It shows the appearance of a slice of amethyst under crossed polarizers

This phenomenon, first noticed by Sir David Brewster in 1815 and described by him in 1823, shows up as orderly lines of rounded bottomed vees. These lines are especially apparent under a microscope in crossed polarized light. Lu Taijing and I. Sunagawa showed that to grow crystals with Brewster fringes, there were two basic requirements, the presence of iron and a solution in which nucleation had started and some crystallites had reached the size of approximately ten nanometers. This situation is typical of the environment in which amethyst crystals grow. What was truly amazing about their work is that under intense magnification the darker streaks turned out to be caused by the dividing lines between right and left hand unit cells. Some of the darker sections consisted of straight line streaks of clusters and in the other direction by zig-zag lines as illustrated in Fig. 23.

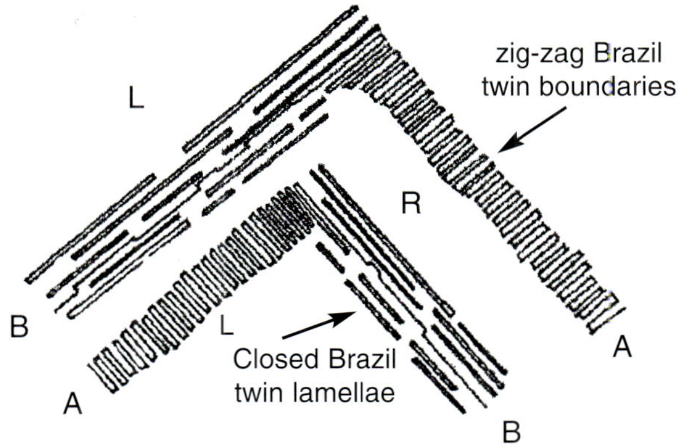

Fig. 23

Double Japan law specimen from Andilamena, Tamatave, Madagascar 3.2 cm high
Tropic Stone Collection Photo by Jeff Scovil ©

Alvin J. Cohen (1989) adds to our understanding of the situation by stating: *"There are channels perpendicular to the major rhombohedral faces that do not occur perpendicular to the minor rhombohedral faces. The presence of iron in these channels produces Brazil twinning to relieve strain in the quartz structure."* In other words the large iron atom is too large for the space in the ordinary unit cell. However, by adding the space in a "left-handed" unit cell to a corresponding space in a "right-hand" cell there is enough room for a Fe^{3+} atom (an atom of iron with three extra electrons) to fit in without forcing aside other atoms and "stressing" the crystal. A left-hand helix of quartz abutting a right-hand helix of quartz is probably the reason for the straight line sections of the Brazil twin boundaries.

Piezo-electric Effect and Commercial Uses of Quartz

One of the unusual features of crystalline quartz is what is known as the "piezo-electric effect" which means that when pressure is applied to opposite sides of a crystal, one side develops a positive electrical charge and the other side a negative charge. This phenomenon was discovered by Pierre Curie and his brother Paul-Jean in 1881. They also found that the opposite effect is true, when electric terminals with one positive and one negative charge are placed on the opposite sides of a crystal, the crystal changes shape slightly. They had no idea of the many important devices to which this purely scientific discovery would lead.

One of the factors that has led to the development of many different devices is the fact that the voltage produced is directly proportional to the pressure and, likewise, the change in the shape of the crystal is in proportion to the voltage applied. This enables special slices of quartz to be used in metering devices of various kinds, including quartz watches and clocks.

Citrine on smoky from Phyllis Ann Claim,
Lake George, CO, USA 3.4 cm high
Crystal Group Collection Photo by Jeff Scovil ©

This phenomenon of the piezo-electric effect works in only certain directions in a crystal and requires uniformly pure quartz. Most natural quartz contains a mixture of right- and left-hand quartz, and slices of mixed quartz will not work. For that reason manufactured crystals are used almost exclusively today.

Devices such as high pressure gauges and sensors measuring pressures in flowing materials often use quartz slices in their construction. But the greatest use of these slices comes from the fact that there is a timed delay between the application of the electrical potential put across the slice and its physical response. By stimulating the slice at regular intervals it is possible to get the slice to vibrate. This vibration occurs at a particular frequency and can be used to set the wavelength of a radio broadcasting or receiving signal. The manufacturing problem is getting the slices cut exactly right so that they will vibrate at the exact frequency desired. These devices are known as oscillators. Almost all timing devices today, such as watches and clocks, use quartz crystal oscillators. Most oscillators for watches are based on crystals vibrating at 32kHz (32 thousand vibrations per second).

It was estimated that, in the year 2000, the United States used about 200 metric tons of synthetic quartz (manufactured using natural quartz to supply the silica needed). The manufacture of synthetic quartz is thus seen to be commercially as well as technologically important.

As is usual with quartz, nothing is simple. The first oscillator was put in use in 1921. Much has been learned about them and their manufacture since then and improvements are still being made. The way slices of quartz crystals move under an electrical stimulus is dependent on the location in the crystal from which it was taken and also the direction in which the slice was cut. The top of the slice may move in relation to the bottom of the slice, the shape of the slice may change, with one dimension increasing and another decreasing, or the length of the slice may change. The speed of vibration of the slice is dependent on its thickness. The frequency of vibration will vary with different temperatures and with the length of time that the crystal has been vibrating. The shape of the slice is also important in determining its characteristics. To prevent unwanted overtones of the vibrational frequency, extreme care is taken with the cutting and polishing or etching of the slices. The exact desired frequency of vibration is sometimes obtained by gold plating the slice with a layer of gold one atom thick, which changes the frequency by 2 one hundred-thousandths of one percent. Computers, television sets and electronic games also use oscillators.

Other characteristics of quartz make it desirable to use for other purposes. Devices like optical lenses and electric lamps also make use of quartz when transparency to ultra-violet light and/or chemical resistance is important. The silicon used for most computer chips comes mostly from quartz which has been processed to remove the oxygen. Sand, which may have begun as a mixture of small particles of various minerals ends up being mainly composed of quartz. The grinding action of wind and water disintegrated most of the other minerals leaving quartz because of its greater durability. When the amount of minerals used commercially is considered on a tonnage basis, sand (and therefore quartz) is, by far, the most important. It is used in the manufacture of glass, concrete and abrasive products like sandpaper.

Specimen from Brazil 9.2 cm wide
Francis Benjamin Collection
Photo by Jeff Scovil ©

The beauty of the color of some specimens, its hardness, as well as its comparatively low price make quartz a most important semi-precious gemstone. The clarity of rock crystal and the richness of the colored varieties: amethyst, rose quartz and citrine ensure its use in many fine pieces of jewelry.

V twin (not Japan law) from Lavra Ariranha, Municipio de Pavai, Teofilo Otoni, Minas Gerais, Brazil
8.2 cm high Herb Obodda Collection formerly in Richard Gaines Collection Photo by Jeff Scovil ©

Quartz specimen from Cinovec, Czech Republic
in the National Museum, Prague Czech Republic 6 cm wide
Photo by Rainer Bode ©

*"... This grand book, the universe, stands continually open to our gaze.
But the book cannot be understood unless one first learns to
comprehend the language and to read the alphabet
of which it is composed."*
Galileo

Color

Color is another topic that shall be mostly skipped over in this publication. Involved is one of the deepest secrets of the universe: the nature of light. Sometimes light acts as a particle and sometimes as a wave. A full explanation of this phenomenon is not yet available and perhaps a mineralogist will some day provide key information leading to the answer. In the meantime, the answers available involve the quantum theory, bands of electrons, the looping orbits of electron pairs and what happens when an atom is missing an electron from one of the pairs. Those of you who are interested may dig deeper into other publications.

Dr. George Rossman (1991) introduces the subject of color in quartz in the following way: "*Quartz, when chemically pure SiO_2, is colorless and transparent to light with wavelengths from about 145 nm in the ultraviolet to about 2700 nm in the infrared when in mm to cm thickness. Natural quartz has three important colored varieties: amethyst (violet), citrine (yellow to brown) and smoky (dark smoky quartz is sometimes called morion). A fourth variety prasiolite (green) also exists. Each of these varieties exhibits either a substitutional or interstitial component other than Si. In addition to chemical substitutions, development of color in some of these varieties requires exposure to ionizing radiation or heat. Each of these varieties can be produced through heat treatment or irradiation of natural quartz of a different color and can be produced by direct synthesis or through synthesis followed by irradiation. Other colored varieties of natural quartz (rose, blue, chrysoprase, jasper, etc.) are, in fact, mixtures of quartz and other phases (although there may be a type of rose quartz from substitutional causes alone).*"

Smoky Quartz

Dr. Rossman also says that the substitution of a few thousand aluminum atoms in a group of a million silicon atoms are sufficient to allow a crystal to be changed from clear to smoky if it is subsequently irradiated. The aluminum atom is larger than the silicon atom and its electrical charge is one less. One less electron is needed from the oxygen atoms to balance the molecules electrically. It is believed that the color is caused by an oxygen atom having one electron missing and thus creating an unbalanced pair. The smoky color can be removed by heating the quartz to 350° C or the color may be increased by passing a current through the quartz when it is at a temperature of 800° C. Color can also be increased by subjecting the crystal to X-rays or gamma rays. A combination of treatments can also turn the crystal into other colors, for example blue, green or yellow-green. It is rather evident that the human eye alone is not able to determine the history of colored quartz.

Rose Quartz

There are two basic forms of rose quartz: crystalline and massive. Clear well formed crystals are quite rare. Few locations have produced crystals of reasonable size. Brazil, Maine, USA and the Kunar range in Afghanistan have produced these crystals. The color is thought to be caused by electrons from phosphorus impurities producing radiation which affects aluminum in the crystal. This color is subject to fading in bright sunlight.

The other form of rose quartz is massive and is found in large quantities in numerous places around the world. The exact cause of the color has not yet been exactly determined. In the 1920's, it was

Tessin shaped crystals with hematite inclusions from 2nd Sovietsky Mine,
Dal'negorsk, Primorsky Kray, Russia 8.2 cm high
George Megerle Collection Photo by Jeff Scovil ©

thought that manganese was the cause of the color. In the 1950's titanium was named as the cause. Needles of rutile have been also considered the reason. Some persons thought that fibers of dumortierite were responsible. In recent tests fibers have been dissolved out of rose quartz by hydrofluoric acid. These fibers have the right color and are the cause of the cloudiness in the quartz. Extensive studies on these fibers have been made by Julia Goreva, Chi Ma and George Rossman, who now (2001) believe that the color is caused by them. The fibers are composed of a mineral fairly similar to dumortierite but not exactly that species. Studies are continuing on this subject and they hope to have the definitive answer soon. The interesting aspect of this report is that while there are large quantities of rose quartz colored by these fibers in certain areas in the world, the cause of the color is a species of mineral apparently found in no other minerals and not enough is known of it to have had the mineral classified so far as a recognized species.

Citrine, Amethyst and Ametrine

After years of wrong guesses there is general agreement that the basic cause of these colors in quartz is iron. The story can never be a simple one because minor amounts of other elements also affect the color. Heat treatments and subjection to radiation will alter the appearance of the stone. Specimens from varying locations are apt to result in different colors when undergoing similar treatments. Bright sunlight will fade amethyst. Another aspect of amethyst is that the color is not uniform within the crystal but is usually concentrated under the major termination faces (the ***r*** faces). That color is also subject to banding (see the section on Brewster's fringes). The details of exactly what happens at each color center have not yet been completely worked out.

Blue Quartz

The blue color is seen only by reflected light and the opposite color, an orange-brown, is seen in light coming through a thin section of the material. The color in blue quartz is caused by inclusions. Ilmenite and rutile are mentioned often as the cause of the color. Dr. Gerhard Niedermayr points out that blue quartz caused by needles of tourmaline has been found in Graubünden, Switzerland, and by needles of crocidolite asbestos in Golling, Salzberg, Austria. Blue quartz caused by aerinite has been found in Antequera, Malaga, Spain.

Green Quartz

The term Prase is used for green microcrystalline quartz, the coloring of which is caused usually by a green amphibole or other green silicates. It has also been used by some persons as a term for crystalline quartz which is so colored. This makes for confusion and it is best to simply use the term green in the same way that the term blue is used.

There are some quartz specimens from the Greek island of Serifos that are unusual. Most of them are colored green by hedenbergite but some have pale amethyst or yellow tints which is caused by iron. Many have hexagonal cross sections but some are cylindrical. Only a few have normal rhombohedral terminations with most coming to a rounded point. The distinguishing feature is the graceful spindle shapes of the crystals, particularly those with the round cross sections. Many are of the “crystal with children” type with the ***c*** axis of the “children” spreading out slightly from the ***c*** axis of the main central crystal.

Quartz group with hedenbergite inclusions
from the island of Seriphos, Greece 8 cm high
Francis Benjamin Collection Photo by Jeff Scovil ©

Specimen from Treasure Mountain Diamond Mine
Little Falls, NY, USA 6.2 cm high
Bob Borofsky Collection Photo by Jeff Scovil ©

Unusual specimen from Brazil 5.8 cm high
Vasconcelos Collection
Photo by Jeff Scovil ©

Graceful duo from Brazil 30.1 cm high
Francis Benjamin Collection
Photo by Jeff Scovil ©

Crystal Shapes

One of the greatest delights of examining quartz crystals derives from the multitude of different shapes that are encountered. Shown here are a few of them.

The simplest form to be found consists mostly of terminations having little or no sides or prism faces. There are three basic types.

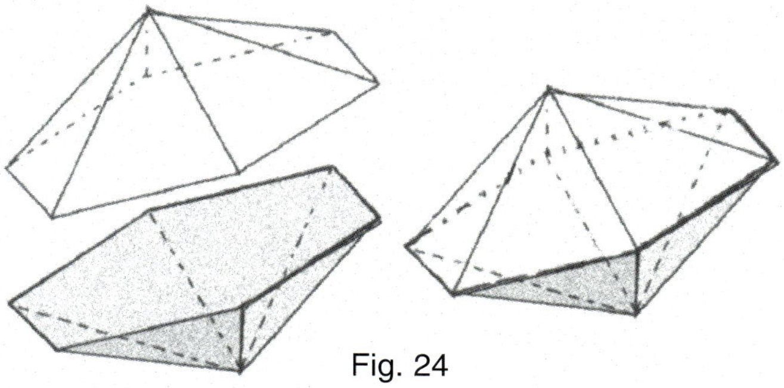

Fig. 24

Some crystals form with twelve faces, in the shape of two six-sided pyramids base to base. These are called hexagonal dipyramids. This shape is typical for β (or beta) quartz but is not common for quartz formed at lower temperatures, α (or alpha quartz).

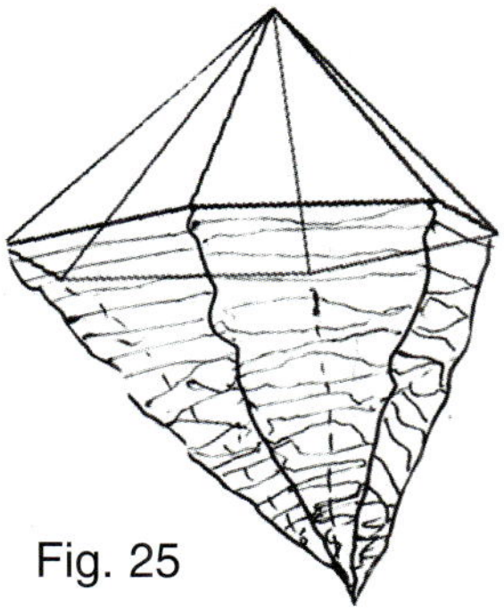

Fig. 25

Not having prism faces applies also to most crystals from geodes where so many crystals are growing so close together. The typical crystal removed from such a formation may have a nice termination but the lower part of the crystal has no particular shape and is reminiscent of the root of a tooth.

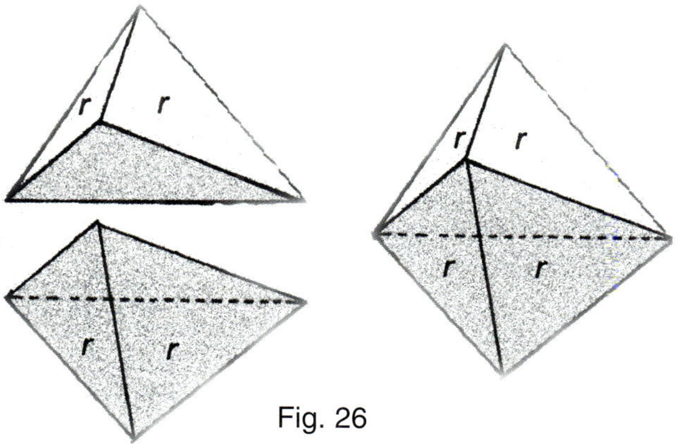

Fig. 26

Some crystals are double terminated and are almost cubic in shape. These are really doubly terminated crystals in which either the ***r*** or the ***z*** faces have grown very large at the expense of the other. The ***m*** or prism faces may be very small or not present at all.

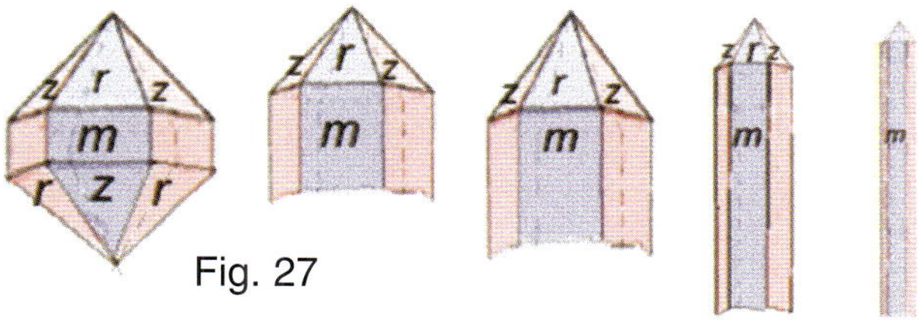

Fig. 27

Illustrated here are several different crystals with different proportions in length and thickness. Because several different factors are at work at one time that affect the crystal shapes, we refer you to the section on growth for a fuller explanation.

Some persons have tried to establish classification schemes based on the ratio of the length of the crystal to its thickness. Because there is no clear dividing point between the classifications, these schemes have generally been ignored.

Cluster from Bourg d'Oisans, Isère, France
6.8 cm high Pascal Bologne Collection
Photo by Jeff Scovil ©

Crystal from Madagascar 5.3 cm high
International Mineral Exchange Collection
Photo by Jeff Scovil ©

"Cathedral" cluster from Characas,
San Luis Potosi, Mexico 3.2 cm high
C & K Graeber Collection
Photo by Jeff Scovil ©

Cluster from Les Pivoirands, Viaille, Isère,
France 4.1 cm wide Eric Asselborn Collection
Photo by Jeff Scovil ©

Types and Forms of Crystals

The exact classification of quartz crystals into types and forms presents many difficulties. This is because there are many ways of distinguishing between specimens and it is easy to create many different categories. One of the questions to be asked is "how many different categories do you want to have?" The features described here for each type will be the ones most typical of that particular type. It should be remembered, however, that crystals grow as a result of their environment and changes in that environment result in a change in the growth pattern of the crystal or crystal cluster. Many different crystal types have some features in common and some specimens really belong in more than one classification. Therefore, some of the specimens illustrated in this book, while they are interesting, unusual pieces, do not fit into one and only one category. Enjoy!

During their "lifetime" many crystals and crystal clusters undergo great changes in their environment. Many of these are permanently recorded for us to "read" from the changes that took place in the growth pattern. The illustration of phantom smoky quartz (on page 74) shows that a whole series of events took place. Sometimes the normally clear solution, in which the crystal was growing, became riled up by the movement of the surrounding rocks or a change in the flow of the solution. Foreign particles of different materials were picked up in one place and carried to the area where the crystals were. Later, when the disturbance decreased, the foreign particles dropped out, some on the quartz crystal. As the quartz crystal grew, it simply enclosed the deposit and kept growing. In the specimen shown it is possible to count several recurrences of disturbance and to note that the intervals between events do not seem to be uniform. Additionally, the composition of the added particles is not always the same.

Group from the Gold Lakes Mine, Silverton, CO, USA 14.6 cm wide
Benjy Kueling Collection Photo by Jeff Scovil ©

In pondering how such events could occur consider how some mountains originate. The Alps, for instance, grew by the addition of large sheets of rocks or "nappes" which were moved hundreds of kilometers from their place of origin to where they are today. The movements were powered by the enormous forces of continental drift and the trip was not smooth. It was instead a "rocky road" that the nappes traveled and during the trip the pressures and temperatures varied and earthquakes could be expected. The composition of the solutions working their way through the rocks could vary from time to time and move in different directions, carrying different ingredients and under varying pressures. There was no uniformity of elements, minerals, gasses or particles in the liquid. Changes could be episodic, that is they could occur or recur many times. Individual specimens must be studied to determine exactly what had happened to them during their long growth period. J. Mullis calculated that quartz in some parts of the Alps had a growth period of over 5 million years.

Scepter quartz with calcite, crystal 3.3 cm high
from East Rock Products Quarry,
St. Johnsonville, NY, USA
W. & V. Haidysz Collection
Photo by Jeff Scovil ©

Beta quartz cluster from 2nd Sovietsky Mine,
Dal'negorsk, Primorsky Kray, Russia
7 cm high
Jeff Scovil Collector and Photographer ©

Smoky quartz from Minas Gerais, Brazil
specimen is 5.6 cm wide
Francis Benjamin Collection
Photo by Jeff Scovil ©

Etched gwindel from Schattig Wichel, Piz Giuf,
Tavetsch, Grisons, Switzerland 7.2 cm high
Eric Asselborn Collection
Photo by Jeff Scovil ©

Tessin crystals with actinolite inclusions from Alchuri, Shigar Valley, Northern Areas Pakistan
12.4 cm high
Francis Benjamin Collection
Photo by Jeff Scovil ©

Rocket Habit Crystal 6 cm high
Blue River deposit,Primorsky Kray, Russia
H. Dibble Collection Photo by Jeff Scovil©

Tapering and Expanding Crystals

Another variation of quartz crystal forms are the crystals that change in thickness as the crystal grows along the ***c*** axis.

Tessin Habit

The type that has a smaller cross section the further the section is from its origin is known as the Tessin Habit for the Swiss canton where it is sometimes found. This type of crystal grew under conditions of high temperature and pressure.

Artichoke or Rocket Quartz Habit

There are some crystals that have the opposite growth pattern and grow larger as they move along the ***c*** axis away from the base of origin. These have been called "Artichoke Quartz" or "Rocket Quartz" and are usually surrounded by secondary crystals.

Tessin habit quartz Becker's Quarry, West Willington CT, USA 6 cm high
Bruce Jarnot Collection Photo by Jeff Scovil ©

Smoky tessin type crystal 15.2 cm high from Hiddenite, NC, USA
H. Dibble Collection Photographer unknown

Muzo and Cipo Habits

One of the most interesting habits of quartz crystals is that which starts from a hexagonal base and grows into a crystal with a triangular cross section. Rudolf Rykart has identified two such habits which he calls Muzo and Cipo. The Muzo type is named for the emerald mine in Colombia, and the Cipo type for the Serra de Cipó, Minas Gerais, Brazil.

The Muzo type has growth striations on the faces under the ***z*** faces of the termination. It would seem that the ***z*** faces grew faster than the ***r*** faces and that much of the growth originated at the edge between the bottom edge of the ***z*** face and the face below it. (Let us designate this as a ***k*** face). As the crystal lengthened, the terminal faces advanced and the ***k*** face grew. The edge between ***z*** and ***k*** lengthened at the expense of the m faces. If the crystal growth had continued long enough, the ***m*** faces would have been eliminated entirely. Note that the striations on the ***k*** face are parallel to the base of the ***m*** face.

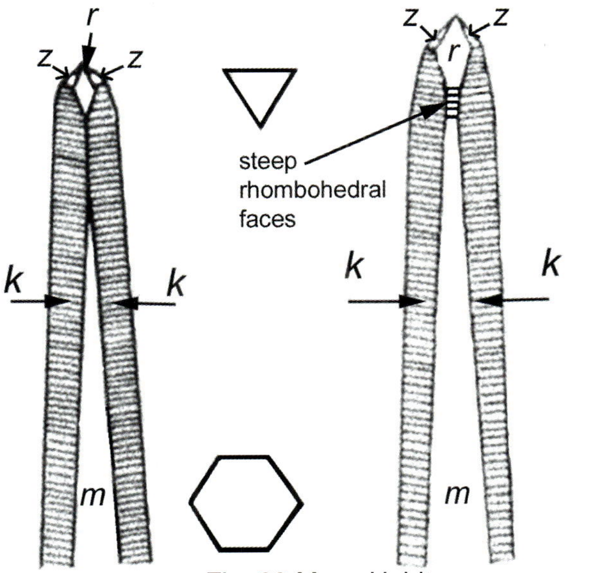

Fig. 28 Muzo Habit

The Cipo type is much the same with the ***k*** faces growing wider at the expense of the ***m*** faces. The difference is that the striations in the Cipo type are parallel to the edge of an ***s*** face. The striations sometimes are parallel to the right ***s*** faces and sometimes to the left counterparts.

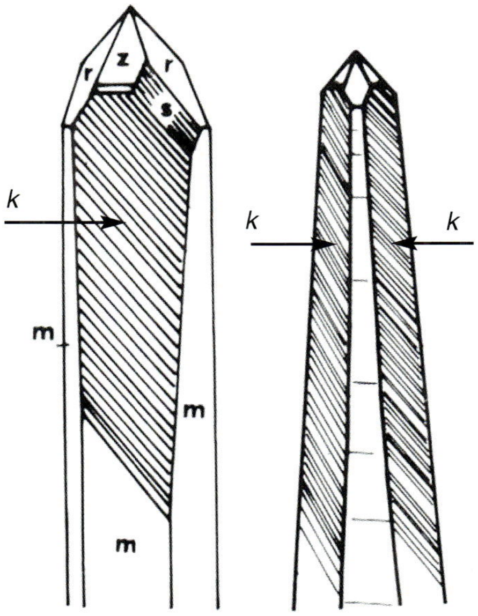

Fig. 29 Cipo Habit

Sketches (Fig. 28 & 29) adapted from Rudolf Rykart: *Quarz-Monographie second edition 1995*

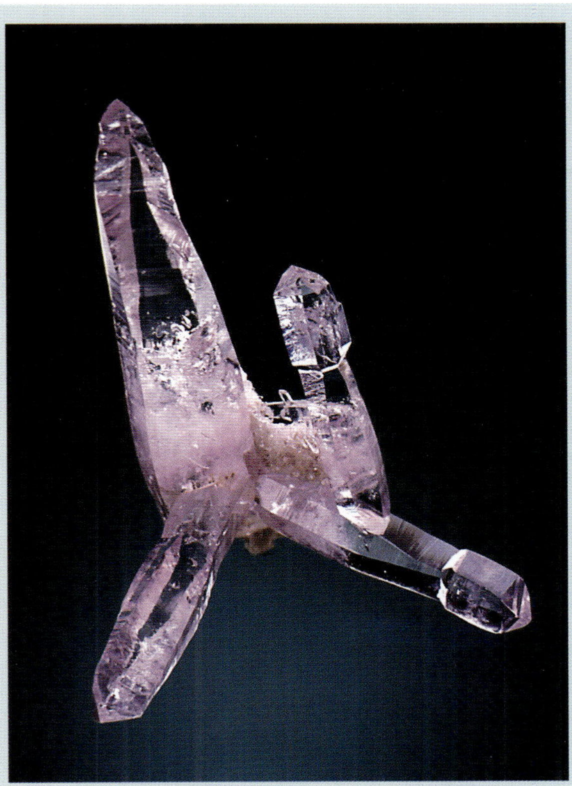

Amethyst scepter crystals from
Las Vigas, Vera Cruz, Mexico
6 cm high
Francis Benjamin Collection
Photo by Jeff Scovil ©

Quartz with actinolite inclusions and with associated andradite from Sinerechenskoye (Blue River) Deposit, Primorsky Kray, Russia cluster 9.7 cm high H. Dibble Collection Photo by Jeff Scovil ©

Faden Quartz

Faden quartz is a type of crystal whose interesting history can be read because of the clarity of the normal crystals. Quartz is designated as "faden" quartz when there is a milky "string" or "thread" that runs through the crystal or crystal aggregate. This phenomenon has intrigued many individuals. Among those who have written about it are Rudolf Rykart in his *Quarz-Monographie* and R. P. Richards in the *Mineralogical Record.*

The scenario of the story of typical faden quartz could be the following:

'A large mass of metamorphic rock is being moved by external forces, possibly the forces of continental drift, so enormous pressure is put on the mass. It buckles slightly at one point in the rock because of the pressure and starts to split and open up a gap. Under high pressure rock is quite porous to liquid solutions and, therefore, a solution with suspended silica and gasses moves quickly in to fill the void. Temporarily the pressure is reduced and the solution moving into the void becomes supersaturated with the silica. Gasses separate out and form small bubbles.

In the material being split is a small quartz crystal which is also parted by the action. The energy in the two parts of the broken crystal quickly attracts the silica particles, lines up their unit cells and builds a bridge between the two parts. During this action some of the gas bubbles and droplets of the solution are trapped by the new growth.

Because the enormous pressures which started the gap remain barely abated this spot in the stony mass which has proven to be a weak spot continues to spread. The quartz crystal splits again and then the entire bridge building process resumes. The series of groups of bubbles of gas and liquid trapped by the bridging process soon looks like a white string with the direction of the "string" the same as the direction that the original parts of the

Faden crystal 8.9 cm high from Darci Ismael, Kahn Dist. Waziristan, Pakistan
Francois Lietard Collection Photo by Jeff Scovil ©

Faden crystal from Bourg d'Oisans, Isère, France
10 cm high Private Collection Photo by Jeff Scovil ©

Faden crystal 7.5 cm high from Darci Ismael, Kahn Dist. Waziristan, Pakistan
Francois Lietard Collection Photo by Jeff Scovil ©

Faden cluster from Dara Ismael Khan Dist. Waziristan, Pakistan 10.2 cm high
Francois Lietard Collection
Photo by Jeff Scovil ©

crystal were moving apart. The faden that forms may be thin like a thread to as wide as a centimeter.'

One of the factors that makes faden quartz interesting is that the "faden" is not always straight. When the original quartz crystal was near the end of the opening gap, it would be as though the sides of the gap were hinged at its end, and the faden would be curved. If there were any movement of the two sides of the gap in relation to each other, that fact would be recorded in the faden. That is the reason that some fadens are zig-zagged or make other unusual lines.

Concerning the tabular second stage growth of the faden aggregate, Dr. Richards writes: "*When the second stage growth begins, it is likely to commence in many places along the faden, creating many nearly parallel crystals, each in effect growing from a single point on the faden. If interrupted at this juvenile stage, the faden crystal is characterized by a relatively thin zone of clear second-growth quartz over the faden, and by many re-entrant angles where the crystal segments originating at different points along the faden have begun to intersect. In the intermediate stage, further growth tends to fill in re-entrant angles and resolve mosaic misfit, producing a faden crystal which looks more like a single crystal, but with some major re-entrants. In the mature stage, this smoothing process is complete, and the crystal is typically tabular but essentially without re-entrant angles. If the growth were to continue indefinitely, the crystal eventually would become very large compared to the faden, the distortion would become insignificant, and the crystal probably would not be recognized as a faden crystal.*"

Rudolf Rykart points out several other minerals which develop line forming aggregates when growing under similar conditions. He names adularia, albite, epidote, augite, diopside, titanite, hessonite, apatite, calcite and hematite.

Crystals with "Children"

There are several crystal habits with smaller associated crystals, bearing various informal names. The difference is in the position of the axes of the smaller adjunct crystals in relation to the axis of the main crystal. These vary from being parallel to a position at 90° to that of the main axis.

Cathedral Quartz

One of the most common of these types is known as Cathedral quartz. These are crystals with the axes of the secondary crystals parallel to those of the main crystal. These clusters vaguely resemble smaller spires or buttresses around the main spire of a cathedral.

Pineapple Quartz

There are crystal groups which have many smaller crystals forming a ring around the main crystal. The axes of these crystals are at approximately right angles to the main crystal. This type is known as Pineapple quartz.

Pineapple quartz from Mineral Farm area Ouray Co., CO, USA 3.2 cm high Tom Rosemeyer Collection Photo by G. Robinson ©

Specimen from Brazil 16.9 cm high Francis Benjamin Collection Photo by Jeff Scovil ©

Not all cathedral quartz specimens are huge. This one from Minas Gerais, Brazil 8 cm high H. Dibble Collection Photo by Sally Dibble ©

Pinetree cluster from Tirniauz, Russia limonite coating field of view 2.8 cm Photo by R. Bode ©

Specimen from the Island of Serophos, Greece is colored by hedenbergite 7 cm high H. Dibble Collection Photo by Sally Dibble ©

Pinetree Quartz

Similar to "Pineapple quartz" but with more pronounced growth of the smaller crystals that surround the base of the main crystal is a type known as "Pinetree quartz". These specimens were found at Tirniauz, Russia by a group organized by the Moscow State University led by L. Nasdala, I. Pekov and Rainer Bode and described in Mineralogical Record, Vol. 27, pp 205-6.

Pinetree cluster on calcite from Tirniauz, Russia field of view 1.3 cm Photo by Rainer Bode ©

Phantom from Shaba, Dem. Rep. of the Congo 7.0 cm high Victor Yount Collection Photo by Jeff Scovil ©

Altered Terminations

There is a group of different crystal types whose growth habits have been changed by a radical change in their environments.

Scepter Quartz

Scepter growth is a knob-like growth on an earlier crystal. What happened was that the solution in which the crystal was growing changed; perhaps by a variation in temperature or pressure or a change in the direction of flow of the solution that resulted in altering its composition. First, there seems to have been a cessation of normal growth. At that time the original crystal became coated with a thin coating of some other mineral. Lately the term Epitaxial Lateral Overgrowth (ELO) has come into use to explain part of what occurs next.

Prof. Ichiro Sunagawa explains the situation as follows: "*After the surface of earlier formed quartz crystals is covered by a thin film of other minerals, like clay minerals, new solution is supplied. Epitaxial growth takes place from an uncovered small portion on the earlier formed crystal surface. Starting from such a portion, epitaxial lateral overgrowth proceeds on the surface of the thin layer of other minerals. This results in the formation of cap portion of scepter quartz, with the same orientation as the earlier formed crystal. ELO mechanism has been shown by us both by investigation of natural samples and hydrothermal synthetic experiments.*"

In most cases the new growth was more rapid than it was originally, due to a higher supersaturation and it proceeded almost equally in all directions. The composition of the new growth was often different from that of the old and the rapid growth led to an incorporation of iron and other elements into the structure. This in turn led to the many fine amethyst scepters to be found on ordinary colorless or smoky prisms. The reverse is also true that some scepters are clear or palely colored on dark stems.

As a result of the different environmental events during growth, one can find such specimens as double scepters and also "inverse" or "reverse" scepters where the new growth is only a fraction of the size of the old. It is a varied and interesting group.

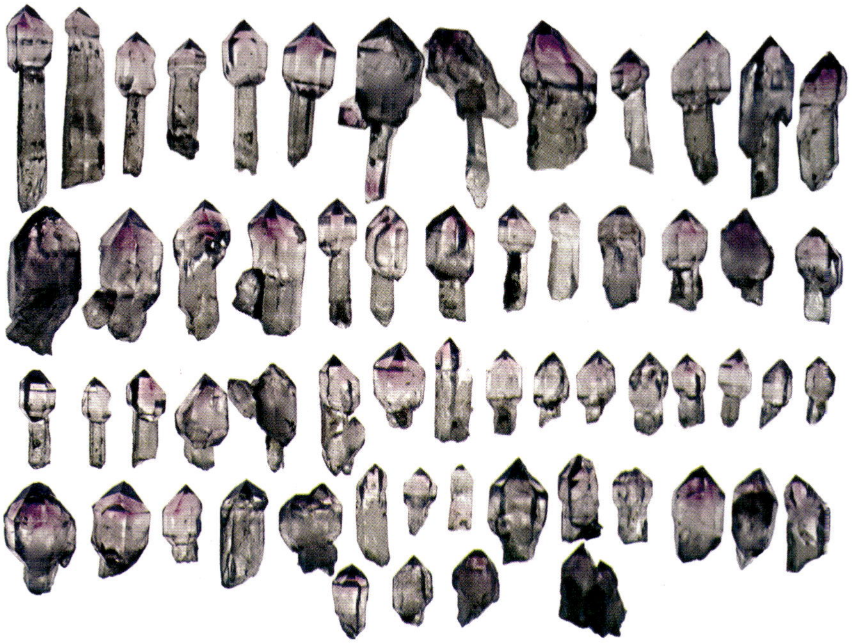

Notice that in many of these crystals the ***c*** axis of the new growth is parallel to but not identical with that of the original crystal.

Partial contents of a single pocket at Fat Jack Mine, Yavapai Co., AZ, USA
Collection of Jeff Scovil Photo by Jeff Scovil ©

Scepter quartz from Treasure Mountain Diamond Mine, Little Falls, NY, USA 10.3 cm high
Bob Borofsky Collection Photo by Jeff Scovil ©

Scepter quartz from Spruce Claim, King Co., WA, USA 5.2 cm high Juergen Tron Collection Photo by Jeff Scovil ©

Scepter from Petersen Mountain, Washoe Co., NV, USA 21.1 cm high Hallelujah Mining Co. Collection Photo by Jeff Scovil ©

Scepter crystal from Petersen Mountain Washoe Co., NV, USA 10 cm high Martin Zinn Collection Photo by Jeff Scovil ©

Amethyst from Statesville, NC, USA 5 x 5 cm A.E. Seaman Mineral Museum, Houghton, MI, USA Gabriel Collection Photo by G. Robinson ©

Window Quartz

The crystal habit known as "window quartz" (or fenster quartz or skeleton quartz) is similar in many ways to scepter quartz in that the windowed crystal is the result of a secondary growth. Window quartz has the additional feature compared to scepter quartz in that the new growth is not normal but has been interfered with. The faces of the new crystal are often incomplete, originating at the edges, forming frames around the "window" and then sometimes growing "panes" over it.

The new growth starts at the edges between faces because of the open sites for the unit cells there and because of the Berg effect. The same environmental changes that caused the supersaturation of the silica would also cause the solution to lose some of its carrying capacity for gasses. Bubbles would form. Carbon dioxide, coming out of the solution, would form small bubbles just the same as it does when a bottle of soda water is uncapped. Methane or other organic gasses if present (and they are likely to have been present in the solution) would also form small bubbles when released. These bubbles could prevent the attachment of the quartz particles to the growing crystal.

Another possible cause of interrupted growth is the build-up on the surface of the faces of particles of clay or clay-like material. This could happen if the solution had been riled and was carrying the particles in suspension.

Each of the thin layers of quartz as they grew from the edges of the crystal toward the center of the crystal face is apt to grow at different rates and by a different amount. The result is unusual patterns of the layers and sometimes "window panes" that enclose the gasses or solids that are there.

Specimen from Bologna, Italy
Apparently grew in a solution with large amounts of tan clay-like material present.
H. Dibble Collection Photo by Sally Dibble ©

The typical "window quartz" crystal shows many layers of growth. This indicates many episodic changes in the environment when the crystal was growing.

Dr. Gerhard Niedermayr points out that studies of the inclusions in specimens from the Eastern Alps show that the outer areas were formed at lower temperatures (and pressures) from the older inner areas, which is exactly as would be predicted.

Amethyst scepter from Andilamena, Tamatave, Madagascar 4 cm wide
Tropic Stone Collection Photo by Jeff Scovil ©

Scepter cluster from Idarado Mine, Telluride, San Miguel Co., USA 1.9 cm high
Dave Bunk Collection Photo by Jeff Scovil ©

Amethyst from Deer Hill, ME, USA
7.3 cm high J.B. Fin Minerals Collection
Photo by Jeff Scovil ©

Some specimens such as this from Wawa, Ontario, Canada should be named hematitic quartz. It has much iron oxide in the crystals which have not been turned amethyst color by natural radiation. 11 cm wide

Inverse Scepter Quartz

Environmental changes which are the inverse of the changes that led to scepter quartz, led to “inverse scepter quartz”. The new growth came from a solution with lower silica saturation and as a result the new growth grew slower, smaller and clearer with fewer inclusions.

Inverse sceptered quartz crystal 3 cm high
from Tafelkop, Gobabisbert Mts., Namibia
H. Dibble Collection Photo by Sally Dibble ©

Skeleton quartz on ilvaite from 2nd Sovietsky Mine, Dal’negorsk, Primorsky Kray, Russia
3 cm high Sam Nasser Collection
Photo by Jeff Scovil ©

Skeleton or Hopper Quartz

This is a type of quartz that overlaps some other definitions. Skeleton quartz is similar to window quartz but does not have smooth overgrowths of crystal faces. Specimens may or may not have smooth central faces. They may or may not be overgrowths of older crystals. The edges between faces are raised but the crystals do not have the true funnellike “hopper” shape that some halite crystals have. Perhaps it can be said you know one when you see one.

Specimen from San Luis Potosi, Mexico
4.7 cm wide
Dave Bunk Minerals Photo by Jeff Scovil ©

Window quartz from Brandberg, Namibia 12.9 cm high
Marshall Sussman Collection Photo by Jeff Scovil ©

Japan law twin from Les Rivoirand, Viselle, Isère, France 3.9 cm wide
Eric Asselborn Collection Photo by Jeff Scovil ©

Japan law twin from Collier Creek Mine, Montgomery Co., AK, USA 10.5 cm wide
Art Smith Collection Photo by Jeff Scovil ©

Japan law twin amethyst from Andilamena, Tamatave, Madagascar 3.2 cm high
Tropic Stone Collection
Photo by Jeff Scovil ©

Double Japan law twin from
Les Rivoirand, Vizille, Isère, France
Eric Asselborn Collection
Photo by Jeff Scovil ©

Japan law twin from Campo Blanco, Ica, Peru 9 cm wide
Martin Zinn Collection Photo by Jeff Scovil ©

Japan Law Habit

The origin of these crystals may be explained by saying that if two small "seed" crystals lie with one of the terminal faces of each almost matching, then what amounts to a pseudo growth step is formed; particles can be attached at this point and growth can begin. This is the situation that accounts for the 84.33° angle found between the two halves of this type of twinned crystal. It was long thought that the growth pattern was a long-sided **V** which developed into a butterfly pattern as shown by Fig. 31.

T. Yasuda and I. Sunagawa made an in-depth study of this particular question and reached the opposite conclusion. The fan shape develops first and then the **V**. If there is growing room, growth then continues in the opposite direction with one side of the crystal becoming dominant and the final stage resembles a **y**.

This type of twinned crystal is interesting because of its vee or butterfly shape and the fact that these crystals often grow much larger than the regular shaped crystals from which they emerge. This type of twinning results in tabular crystals with an angle of 84.33° between the c axes. The two sides have the same surface plane even when there is a groove between the two halves. Japan law twins are growth twins and not intergrowth twins. This is because the two parts grow side by side with a common boundary and are not mingled or intermeshed.

Japan law twin with smoky color zones from Furious Titan Claim, El Capitan Mtn., Lincoln Co., NM, USA 2.7 cm wide
H. Dibble Collection Photo by Sally Dibble ©

Frenchmen will point out that this type of crystal was first found at La Gardette, France, and should be named for that location, but later similar crystals were found in greater abundance and size and in more than one location in Japan. Detailed scientific studies on these specimens from Japan brought interest and publicity, and they became known as Japan law twins around the rest of the world.

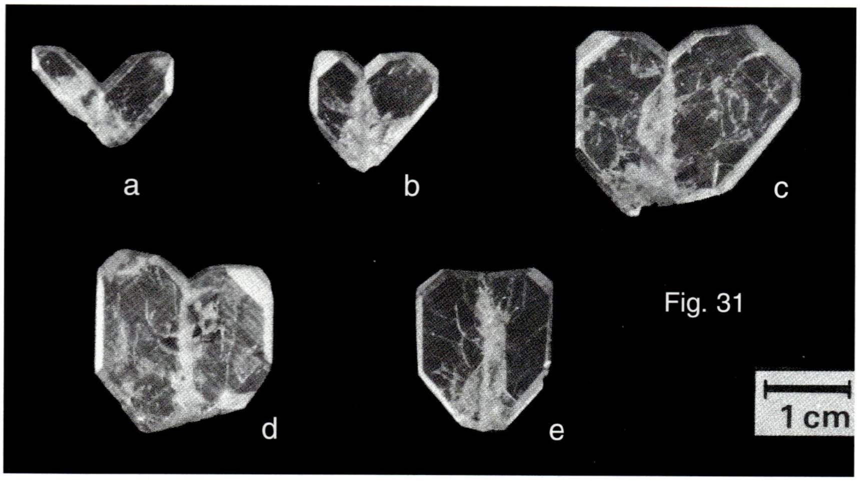

Morphological evolution of Japanese twin as growth proceeds (a to e) expected if re-entrant corner effect is operative.
Yasuda and Sunagawa

Specimen "A" from Dal'negorsk, an example of interrupted growth 12 cm high
H. Dibble Collection Photo by Sally Dibble ©

Specimen "B" from Dal'negorsk, an example of interrupted growth 11.8 cm high
H. Dibble Collection Photo by Sally Dibble ©

Reverse side of Specimen "A" from Dal'negorsk

Interrupted growth cluster from Dal'negorsk, Primorsky Kray, Russia Largest crystal 6.0 cm high
H. Dibble Collection Photo by Sally Dibble ©

Interrupted Crystals

Some of the most unusual forms of crystals are caused by interrupted growth mechanisms. Window quartz is caused by bubbles or particles of foreign matter on the faces of crystals but there is another type of obstruction of growth that too causes very unusual crystals. It is the presence of crystals of other mineral species blocking the places where quartz is attempting to grow. If the foreign matter is not too large, the quartz simply engulfs it and moves on, but if it is larger it reacts in different ways.

One of the most unusual groups of specimens whose growth has been interfered with are from the Bor quarry of Dal'negorsk, Russia. There are changes in the growth forms of the quartz which are varied and interesting. These are in response to partial blockage by calcite crystals as thin as sheets. At times the magnetic attraction of the quartz operates through the calcite and the sections on both sides of the calcite correspond. Sometimes, possibly due to variation of flow of the supersaturated solution around the quartz, the crystal grows larger in one direction than in another. At some stage in the growth of the quartz, the calcite disappeared, leaving only the quartz. The calcite disappeared from the quartz but did not disappear from other minerals in the same quarry. We have included here a photograph of a datolite cluster which shows the sheetlike calcite crystals in close array in the same way that they apparently were in the quartz before they disappeared.

It is rather humorous that some of the people living in the area of the quarries did not understand the cause of the unusual shapes. They thought that the unusual forms were the result of a series of breaks caused by earthquakes.

Datolite and calcite from Dal'negorsk, Russia
specimen 5 cm wide
H. Dibble Collection Photo by Sally Dibble ©

Quartz from Dal'negorsk, Primorsky Kray, Russia
11.5 cm high H. Dibble Collection
Photo by Sally Dibble ©

Skeletal crystal from Gold Lakes Mine,
Silverton, CO, USA
7.9 cm high Benjy Kuehling Collection
Photo by Jeff Scovil ©
Note crystallites in centers of faces

Scepter crystal from Petersen Mountain,
Washoe Co., NV, USA 10 cm high
Martin Zinn Collection Photo by Jeff Scovil ©

Amethyst from Branderg, Namibia
7 cm high
Eric Asselborn Collection
Photo by Jeff Scovil ©

Skeletal crystal from Gold Lakes Mine,
Silverton, CO, USA 7.8 cm high
Benjy Kuehling Collection
Photo by Jeff Scovil ©

Quartz "*Diamonds*"

Rudolph Rykart refers to this group as "Strongly shining small crystals in cavities in sedimentary rocks." These crystals, because their faces are so smooth, look as though they had been faceted and are referred to as "diamonds", usually preceded by the name of the locality where they were found, i.e. Herkimer (NY State,USA), Marmarosh (Ukraine), Mirabeau (France), Oviedo (Spain), etc. Although they have been found in several different countries they have certain common characteristics. They are generally short. They are either scepter growths on an older stem or they grow like the new growth section of scepter crystals, having almost equal dimensions. They are found in small vugs in limestone or dolostone (which has both calcium and magnesium present) and seldom grow larger than four centimeters in diameter. These crystals are often exceptionally clear and they have few faces. The faces are usually limited to ***r, z, m,*** and ***s*** faces but corresponding faces are not uniform in size with each face apparently growing at a different rate.

The experts are still not in agreement as to what the exact sequences were that led to the formation of these crystals but we know in general what happened, particularly around the Herkimer, NY area. There the dolostone in which the quartz crystals formed, was deposited at the bottom of a shallow sea in the Cambrian or Ordovician era approximately 500 million years ago. In that sea were coral formations made up mostly of calcium carbonate, and probably some magnesium. Also present were various organisms and other organic materials (from bacteria on up) including stromatolites. These stromatolites were a form of algae which was layered like cabbages. As these organisms died they settled to the bottom of the sea and deposits slowly built up. Apparently the sea bottom sank slowly and the sea remained shallow. It is estimated that the buildup on the sea bottom took place over about 200 million years and reached a depth of approximately five to seven kilometers. The temperature in the depths rose slowly due to the heat of the earth and the pressure. The organic compounds altered into hydrocarbons

"Herkimer diamonds" from Herkimer, NY, USA
5.2 & 4.6 cm high
Keith Proctor Collection Photo by Jeff Scovil ©

including gasses, liquids and solids, particularly anthraxolite. The temperatures approached but probably did not reach 200° C. The reason for this estimate is the fact that if the temperature had reached approximately 200° C, the dolomitic limestone would have metamorphosed into marble.

When the organic materials disintegrated, due to this temperature and pressure, some formed acids that dissolved the dolostone and created small pockets or vugs in the porous stone. Another factor came into play; some organic compounds have the ability to increase the solubility of quartz in water and that is what apparently happened here. The organic materials in the deposit were not uniformly distributed. For that reason these vugs are quite stratified and limited in area. In these pockets quartz separated out of the underground solutions. The growth of the crystals was slow because there was little change in temperature, the amount of silica in suspension was low and the units that attached themselves to the growing crystals were small. This slow growth was the cause of the clarity of these crystals and the smooth faceted appearance of the faces.

There is still considerable variety in the specimens. For example, some are sceptered on clear stems and some are on black stems. There are a variety of inclusions in the crystals; in addition to hydrocarbons (gas, liquid and solid) some have brine, and some pyrite or other minerals. The fact that there were scepter crystals formed and particularly the fact that often the composition of the scepter growth differed from that of the original crystal, indicates that there were "episodes" during the creation of these crystals. Changes took place in the environment of the crystals. Conditions were not uniform over the entire area. Different crystals may be found in different locations and at different elevations. Trying to figure out exactly what happened is what makes the study of these crystals so interesting.

"Herkimer diamond" cluster 5.0 cm wide from
Middleville, NY, USA
Canadian Museum of Nature Collection
Photo by Jeff Scovil ©

"Herkimer diamond" from Ace of Diamonds Mine,
Middleville, NY USA 8.0 cm high
Francis Benjamin Collection
Photo by Jeff Scovil ©

Quartz scepters 2.5 cm high from Sauberg Mine,Ehrenfriedersdorf, Saxony, Germany
Photo by Rainer Bode ©

David McCullough in his biography of John Adams points out that in 1756 John wrote in his diary to the effect that God created some "wondrous works" in nature but that his greatest gift to man was an inquiring mind.

Two views are shown of this open gwindel cluster to better illustrate the twist between the individual crystals. It is 2.3 cm high and comes from the northern Ural Mountains, Russia
Photo by Dibble Trust Studio ©

Three pieces of quartz from Thunder Bay Area Ontario, Canada showing how a little difference in iron content can change the color of the specimen.
Photo by Dibble Trust Studio ©

Gwindel Quartz

Gwindel clusters are one of the most interesting and unusual of all quartz types. Gwindel quartz clusters are difficult to describe. One way to think of the appearance is to consider a stack of crystals made with the ***c***-axis of the individuals in parallel and then subjected to a twist. The separateness of the individual crystals will depend on the stage of growth of the cluster.

R. Rykart says that '*gwindels start when a gap is opened up in a rocky mass, the same as the gap that starts faden quartz growth. The original crystal lies along one side of the gap with its axis parallel to the slit. The growth then proceeds in the direction of one of the* ***a****-axes of that crystal, toward the center of the gap. The clusters are usually either mostly left or right hand quartz with some Dauphiné twinning sometimes present. Looking down on a gwindel along the common* ***a****-axis, those clusters composed of left hand quartz will have a right hand twist and those composed of right hand quartz will have a left hand twist. The amount of twist can vary considerably with 1 to 5 degrees per centimeter being normal. The thicker clusters tend to have the least twist.*'

In some specimens individual crystals are very distinct but in others the space between the crystals fills in and the sides of the cluster form a fairly smooth, warped tabular surface. Clusters of this sort are called "closed". Those with clearly defined individual crystals are "open" and those in between are "semi-closed".

Gwindel clusters are quite rare and usually the other crystals, which are found near the gwindel in an alpine cleft, are normal crystals. The amount that the local environment affected the creation of this type of crystals during their time of growth is seen from the fact that in Switzerland they are mostly found in the northern Alps, are missing in the central Alps and rare in the eastern Alps.

M. Fehlmann of the Laboratory of Crystallography in Zurich, Switzerland, and A. F. Moodie of the Royal Melbourne Institute of Technology in Melbourne, Australia, pointed out in 1992 that although scientists had been searching for more than 100 years they had not been able to ascertain the cause of the twisting in "gwindel quartzes". Fehlmann and Moodie used several different methods in their study in addition to the usual tests. They used X-ray Laue diffraction and confirmed that the twisting arises at microscopic levels. They used white beam synchrotron radiation topography to prove that neither twinning nor dislocations were responsible. They tried electron microscopy, 'selected area diffraction' and large angle convergent beam micrographs as well as conventional convergent beam electron diffraction. Their conclusion was then "We do not know the cause of the twisting." Different possible causes have been proposed since they presented their paper, but Dr. Fehlmann recently reconfirmed the conclusion they came to ten years ago to the effect that we still do not know the cause of the twisting.

Quartz with fluoroapatite 4.0 cm high
Pulsifer Quarry, Mt. Apatite, ME, USA
George Holloway Collection
Photo by Jeff Scovil ©

Closed gwindel from Hinterer, Ziggenstock, Berne, Switzerland
4 cm high Eric Asselborn Collection
Photo by Jeff Scovil ©

Gwindel crystal in group from Puiva, Tyumen Oblast, Russia
7.6 cm wide
Richard Kosnar Collection Photo by Jeff Scovil ©

Gwindel quartz from Col de Cristaux, Chamonix, Haute - Savoie, France 14.4 cm wide
Eric Asselborn Collection Photo by Jeff Scovil ©

Gwindel quartz from Ziggenstock, Switzerland 12.2 cm high
Eric Asselborn Collection
Photo by Jeff Scovil ©

Gwindel half open - half closed from Kreuzlistock,
Tavetsch, Grisons, Switzerland 6cm high
Eric Asselborn Collection Photo by Jeff Scovil ©

Closed gwindel from Giuv glacier, Tavetsch,
Grisons, Switzerland 4.8 cm high
Eric Asselborn Collection Photo by Jeff Scovil ©

Closed gwindel from Cavradi, Graubünden,
Switzerland 2.8 cm high Ron Pellar Collection
Photo by Jeff Scovil ©

Gwindel from Dodo Mine,
Tyumen Oblast, Russia
6 cm high Van Scriver Collection
Photo by Jeff Scovil ©

Gwindel from the Dodo Mine,Tyumen Oblast, Russia 5.8 cm high
Van Scriver Collection Photo by Jeff Scovil ©

Cluster with hematite and other inclusions
from Dal'negorsk, Primorsky Kray, Russia
4.8 cm high Jeff & Gloria's Minerals
Photo by Jeff Scovil ©

Quartz with rutile and hematite inclusions
9.5 cm high from Novo Horizonte, Brazil
Shirley Fiske Collection
Photo by Jeff Scovil ©

Inclusions

Quartz crystals that include other minerals in them form one of the largest and most interesting of the groups. The number of different species that have been surrounded by quartz must number in the hundreds. Because quartz is one of the last minerals to crystallize, it simply engulfs crystals of other minerals in itself. The lovely colors of many quartz crystals are due to the inclusions of microcrystals of other minerals. For example the soft grayish green of prase is due to actinolite and the color of some blue quartz is due to ilmenite. Particles of hematite cause the coral pink color of crystals from Dal'negorsk, Russia. Small crystals of mica cause the sparkle in adventurine. Quartz with rutile or tourmaline crystals can be very attractive because the enclosed crystals are bright and arranged at all angles in the quartz. When crystals of the lead ore, galena are embedded in the quartz they do not become dull from oxidation and have an amazing luster. This list could go on and on, but these names should be enough to indicate the extent of the variety that exists.

Rutile inclusions in quartz crystal
6.0 cm high from Brazil
Palph Mueller Collection
Photo by Jeff Scovil ©

Faden cluster from Dara Ismael Khan District, Waziristan, Pakistan 13.5 cm high
Collection of the Houston Museum of Natural Science Photo by Jeff Scovil ©

Phantom Crystals

The list of different species of minerals that have been engulfed by the growth of quartz crystals is a long one. Many of the specimens are interesting and some are beautiful. Sometimes other minerals or fluids settle on the faces of a growing crystal. The quartz crystal keeps on growing and includes them as a "ghostly" outline inside the crystal. When an observer is able to discern the shape of the crystal at a previous stage of its growth, the specimen is known as a "phantom" crystal. It is possible to learn much of how it grew from the inclusions. The changes that took place as the crystal grew indicate the changes that took place in its environment. The fact that the crystal did not grow uniformly in all directions and the changes that took place in the layers become a challenge to the observer to figure out its history.

Phantom crystal 4.5 cm high from
Minas Gerais, Brazil H. Dibble Collection
Photo by Sally Dibble ©

Quartz with hematite inclusions from China
3.2 cm high
Wydar Collection Photo by Jeff Scovil ©

Smoky crystal with chlorite phantoms
10.5 cm high from Presidente Kubitschek,
Minas Gerais, Brazil
H. Dibble Collection Photo by Sally Dibble ©

Flattened parallel growth cluster from Brazil 16 cm wide H. Dibble collection Photo by Sally Dibble ©

Péle de Jacaré or Crocodile

Péle de Jacaré or Crocodile quartz is the name given to crystal clusters that look like scales of crocodiles. These clusters, which are found most often in Brazil, are of smoky quartz. The black "smoke" is not uniform in the crystals but is in certain areas, mostly at the edges of the crystals. The axes of the crystals making up the cluster are for the most part parallel. The crystals are short and the prism faces that make up most of the surface of the cluster are fairly smooth.

Crocodile quartz cluster 13 cm wide from Minas Gerais, Brazil H. Dibble Collection

Clusters of Tabular Crystals

Tabular crystals are to be expected with Japan law twinning and faden quartz because of the nature of their formation, but it is not clear why tabular shape becomes the dominant growth habit in some single crystals and in some jumbled clusters.

Details of tabular cluster from Brazil showing flattened parallel side crystals.

Jasckstraw cluster
6.1 cm wide
From Jeffrey Quarry, Little Rock, AK, USA
H. Dibble Collection
Photo by Sally Dibble ©

Haystack or Jackstraw Quartz

One peculiar growth habit is that of "Haystack Quartz", the most famous location for which is the Jeffrey Quarry near Little Rock, Arkansas, USA. Crystals from this location grew completely at random. Many of the crystals are multi-terminated. Clusters are composed of crystals which are stacked every which way like sticks from the game "Jackstraws". Most have milky centers. The crystals formed in rectorite clay during the Permian period approximately 250 million years ago.

Amethyst from the Anahi Mine, Puerto Suarez, Santa Cruz, Bolivia
3.5 cm wide
Houston Museum of Natural Science Collection
Photo by Jeff Scovil ©

The crystals rise from a bed of white quartz and have dark tips indicating a change in the solution during growth. Probably an increase of iron in the solution.

Globular quartz aggregation
5 cm high from Nazik, India
H. Dibble Collection
Photo by Sally Dibble ©

Microscopic detail of the edge of one of the clusters

Globular Quartz

One of the most recent additions to the list of unusual quartz habits is one in which the forms make roughly globular clusters. This habit was described first in Rocks and Minerals Magazine of November/December 2001 by John S. White.

These clusters may grow independently or in turn unite into larger aggregations. Where exactly these are found is not clear. The site/sites are given as in the Maharashtra Province of India, in either the Nasik or Jalgaon district or in both districts.

They are quite clear inside with a generally pebbly frosty appearance. Dealers had been listing such specimens as chalcedony until John took a close look at some of them. He said, *"All one need do is hold a specimen of this material under a light to see countless reflections coming off of tiny crystals faces over the entire quartz surface. This would not happen if this truly were the chalcedony variety of quartz, which is seldom transparent."* He had a thin section made. When viewed under a microscope, it proved that these clusters consist primarily of a group of crystals growing radially from a center or from the surface of the brown matrix.

Smoky cluster from Phyllis Ann Claim, Lake George, Teller Co., CO, USA circa 20 cm high
George Fisher Collectiion Photo by Jeff Scovil ©

Irregular tiered edges on this specimen from
Minas Gerais, Brazil 8 cm wide
H. Dibble Collection Photo by Sally Dibble ©

Detail of a section of the surface of a crystal
from Minas Gerais, Brazil

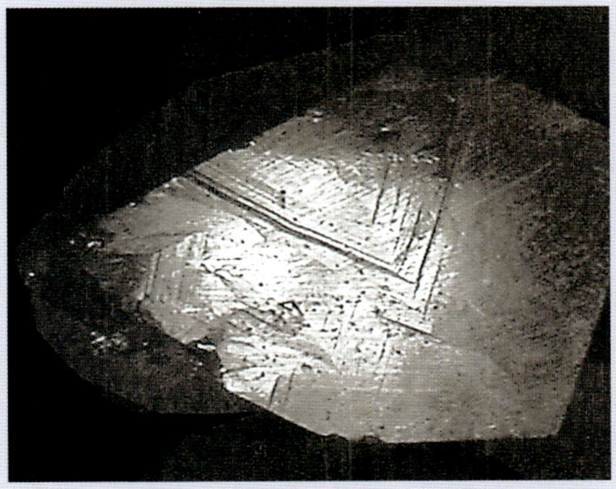

Detail of a section of the surface of a crystal
from Minas Gerais, Brazil
Photo by Dibble Trust Studio ©

Quartz clusters appear in some unusual places.
This cluster is 1.0 cm wide from the
Dugway Range, Juab Co UT, USA
H. Dibble Collection Photo by Sally Dibble ©

Smokey and amethyst crocodile quartz
8 cm high
from Galiléia, Minas Gerais, Brazil
H. Dibble Collection Photo by Sally Dibble ©

Plates of Tiered Crystals

Another unusual crystal cluster type can best be described as a tiered arrangement of tabular crystals with the axes of the crystals in parallel. Two axes at 90º to each other dominate the arrangement and it is not immediately apparent which is the ***c*** axis.

Tiered cluster of flattened crystals from Galiléia Minas Gerais, Brazil 9.0 cm wide
H. Dibble Collection Photo by Sally Dibble ©

"Odd Balls"

There are some specimens of quartz that are difficult to put into any category. For want of a better name the slang term "odd balls" is being used. Following are a few special pieces.

Parallelogram

Because of its cross section this name was chosen. The example chosen is from Minas Gerais, Brazil. It is doubly terminated with multiple points at each end. Overall it is 23.4 cms long and the measurements of the sides, taken at the middle, are 2.1, 1.0, 1.6, 2.2, 0.5 and 1.6 cms. Instead of the ordinary hexagonal cross section with equal width sides, two of the sides are reduced to a minimum and two have grown. This seems to be a case which illustrates the fact that much growth originates at the edges at the base of the termination. Once this line is established as the main center of growth, new layers are added continuously. That would account for the repeated disproportionate shape and the presence of the striations on all surfaces that would normally be smooth prism faces.

"Parallelogram" 23.7 cm long from Brazil
H. Dibble Collection

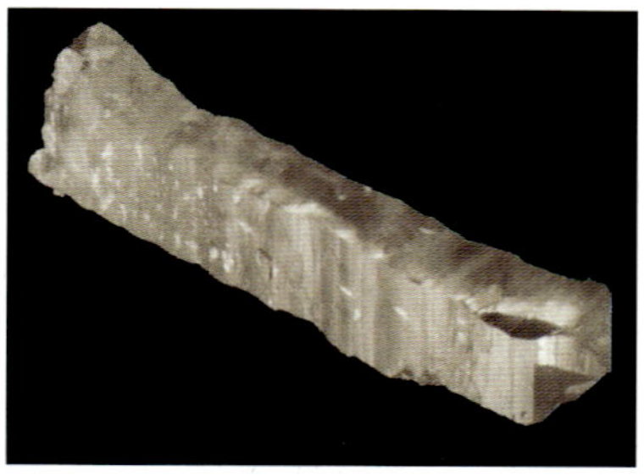

Angle view of "parallelogram"

This specimen from the White Queen Mine in Pala, CA, USA was carved by Susan Allen and shows iris nicely. Photo by Dibble Trust Studio ©

Iris or Rainbow Quartz

Quartz which has developed internal cracks due to stress shows rainbow effects due to the interference of light and is referred to as "iris" or "rainbow" quartz. The iridescent effect is caused by the same phenomena as the colors seen on a film of oil floating on water. To understand exactly what happens requires an understanding of how light sometimes acts as a wave, the different wavelengths of visible light and other physical matters. Because of the room that an explanation would take up, you are referred to other publications such as college textbooks on physics, encyclopedias and good search engines for the internet. Recently *Google* turned up 153,000 references to an inquiry about "light - interference - color".

Amethyst cluster from Irai, Rio Grande do Sul, Brazil 18 cm high
H. Dibble Collection Photo by Sally Dibble ©

Hemimorphic Type Crystals

This unusual crystal is double terminated and has two sections which are different from each other. One section (shown here at the bottom of the crystal) has the usual hexagonal cross section. Then approximately 5.7 cm from the end begins a "Muzo" crystal which is about 10.5 cm long. The ***c*** axes of the two sections form a slight angle to each other.

Left side view Right side view

Specimen from Brazil

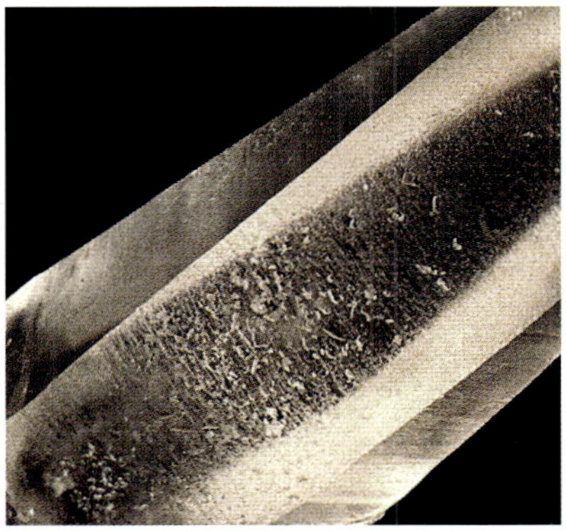

This detail of the face of the hemimorphic crystal indicates a supersaturated solution and fast growth

From Mogilata, the Madan dist. Bulgaria

Bilateral Parallel Quartz Druses

Dr. Ivan Bonev points out another growth habit of quartz, one that he calls *bilateral parallel quartz druses* and provides two photographs from Mogilata to illustrate his point. He uses the definition of *druse* as "crystal growth in a cavity which results in numerous crystal tipped surfaces".

The geological history of the area is reconstructed to be as follows. First some of the original minerals dissolved out leaving pockets. Into these pockets paper thin calcite crystals grew. Then rising hot solutions rich in silica provided the quartz which grew epitaxially on the calcite. The increasing acidity of the solutions dissolved away the calcite which was filled in with quartz particles.

From Mogilata, the Madan dist. Bulgaria

Try to read the history of this specimen from Shan Bao, Hunan, China 7.8 cm wide
Collector's Edge Collection Photo by Jeff Scovil ©

Cross-sections of amethyst statactites 7.0 and 5.2 cm wide from Artigas Uruguay Rock Currier Collection
Photo by Jeff Scovil ©

This Fabre Mineral specimen is from Chelva, Valencia, Spain 2.5 cm wide Photo by Francesc & Jordi Fabre ©

Stalactitic group with goethite inclusions 11 cm high from Tizirine, Oudrzazak, Morocco.
H. Dibble Collection
Photo by Francesc & Jordi Fabre ©

This 19 cm wide specimen from Brazil has gone through some changes in its environment. It looks as though clear chips with smoky edges were inserted into the original lumpy white crystals. Photo by Dibble Trust Studio ©.

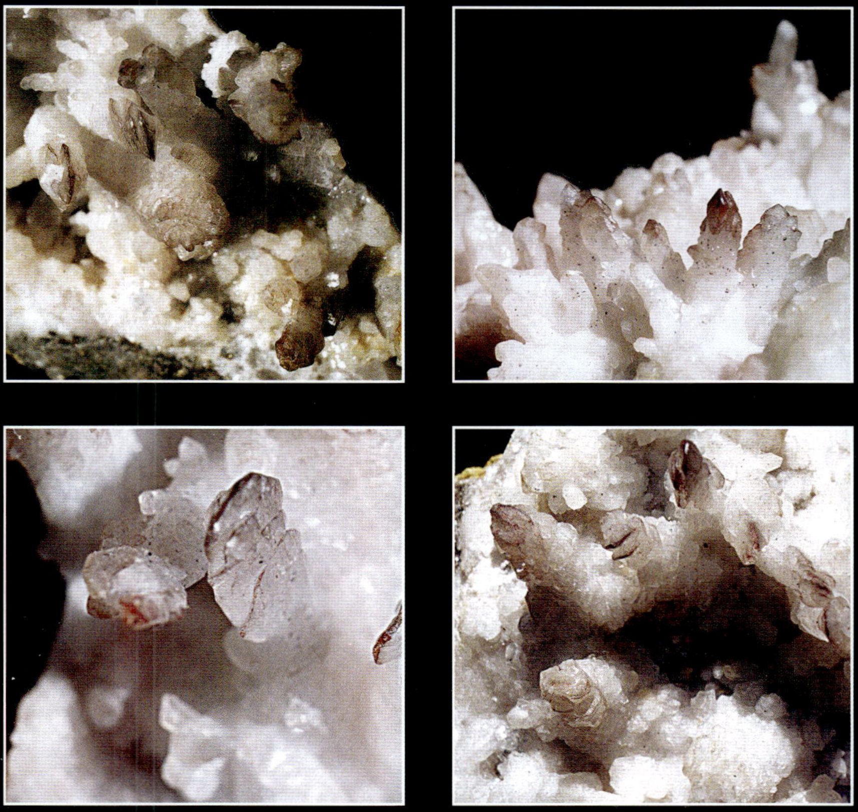

Pictured above are close-ups of the specimen.

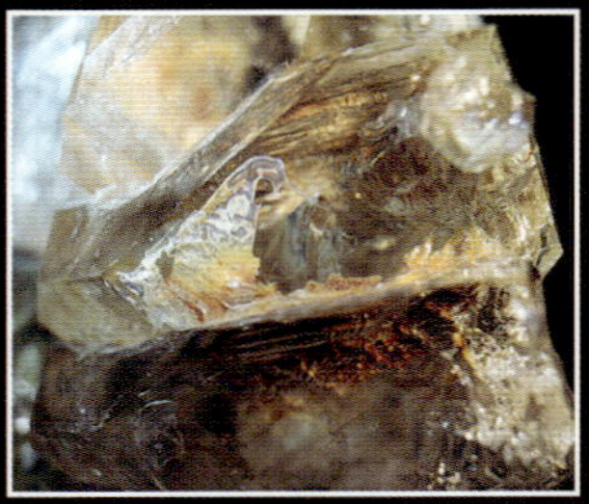

Pictured above are close-ups of the window quartz.

Some specimens have been through so many different episodes in their history that exactly what happened is not clear. Using a good means of magnification, a study of various aspects and views of the piece will show designs and sculptures that by themselves could be classified as art. This specimen of smoky quartz with clay inclusions which is 16.5 cm high from Coronel Murta, Minas Gerais, Brazil is one example. The following pictures are photomicrographs of some of its various aspects. Photo by Dibble Trust Fund Studio ©

Pictured above are photomicrographs of the window quartz on page 86.

This specimen of smoky window quartz with clay inclusions is another example of a piece with a complicated history. 14 cm high
Boca Rica Mine, Galiléia, Minas Gerais, Brazil

Pictured are photomicrographs of the above window quartz.

This is a novelty item faceted by Arthur Grant. It was cut so that a black tourmaline crystal is at the center and perpendicular to the face of the stone. Reflections off the facets give it this startling appearance. 6.01 carats from Brazil

Paper weight by Arthur Grant and Susan Allen
5 cm wide H. Dibble Collection
Photo by Jeff Scovil ©

Ametrine 60 ct square cut in Brazil.
A beautiful stone but too large for a "pinkie" ring.
In a private collection. Photo by Jeff Scovil ©

Gemstones and Novelty Items

Since prehistoric times man has valued quartz and worked pieces of it to enhance its beauty and to produce practical and unusual items. The Greeks thought that wine drunk from a cup fashioned out of amethyst was non-intoxicating. For that reason they valued highly any such cup.

Quartz has been a favorite mineral for faceters because of its beauty and relatively low price. Some very large specimens have been faceted. The Smithsonian Gem Collection, for example, contains amethyst, citrine and smoky faceted stones of over 1 ,000 carats each. It also contains a clear faceted egg shaped specimen of 7,478 carats.

Because images reflect off of the facets, inclusions in specimens can produce very unusual appearing pieces.

Quartz has been used by sculptors to carve both internal and external pieces of great beauty. Novelty items both carved and faceted have been made that show great imagination.

Smoky quartz from Brazil 124.2 cts Portuguese cut
Faceted by Arthur Grant
In a private collection. Photo by Jeff Scovil ©

A phantom crystal and a phantom Mayan priest.
Intaglio by Susan Allan 5.5 cm high from Brazil.
In a private collection. Photo by Jeff Scovil ©

"The Smile" by Arthur Grant
41.77 cts 2 cm wide
Half round - half faceted
1/4 dull 3/4 polished An amazing piece.
H. Dibble Collection Photo by Jeff Scovil ©

Citrine quartz from Brazil.
35.89 cts barion cut.
In a private collection.
Faceted by Arthur Grant.
Photo by Jeff Scovil ©

"Hummingbird" by Susan Allen
513 cm wide reverse sculpture in Brazilian quartz with inclusion.
H. Dibble Collection Photo by Jeff Scovil ©

Quartz piece 4.1 cm wide with internal sculpture by Susan Allen
In a private collection Photo by Jeff Scovil ©

Reverse sculpture by Susan Allen in amethyst stand by Michael Christie in charoite and gold 3.8 cm wide
Now in a private collection. Photo by Jeff Scovil ©

Faceted clear quartz oval cut stone
714 carats from Brazil Faceted by
John Bradshaw Photo by Sally Dibble ©

Smoky quartz 121.5 cts from Brazil.
Custom cut faceted by Arthur Grant.
In a private collection. Photo by Jeff Scovil ©

Citrine from Brazil 35.89 cts.
Faceted by Arthur Grant
In a private collection Photo by Sally Dibble ©

Rose Quartz 43.42 cts Slightly cloudy
but beautiful for rose quartz.
Faceted by Arthur Grant
H. Dibble Collection Photo by Jeff Scovil ©

Tourmaline crystal included in quartz
from Brazil 9 cm high
Arthur Grant Collection Photo by Jeff Scovil ©

Crystal Face Surfaces

The surface textures of quartz crystals vary widely and a study of these features should not be neglected. Most surfaces, unless the crystal grew very slowly, are not perfectly smooth but have interesting designs. Much can be learned concerning the conditions of growth through a careful examination. Turbulence in the growth pocket, direction of flow, if any, of the surrounding solution, rise in the temperature, etc. can sometimes be determined by the appearance of the crystal.

A good magnification device and illumination from the side can lead to an enjoyable time and a better understanding of the events that occurred during the growth of the specimen. Here are a few pictures of different surfaces.

Smoky quartz floater of unusual texture from Perena, Salamanca, Spain 5.2 cm wide
H. Dibble Collection Photo by Sally Dibble©

Photos by Dibble Trust Fund Studio ©

Properties of Alpha Quartz	
Class	Silicates
Sub-class	Tectosilicates
Group	Silica
Chemical formula	SiO_2
Crystal system	Hexagonal
Crystal class	Point group 32 Trigonal-trapezohedral
Space group	α quartz: $P3_121$ (left) or $P3_221$ (right) β quartz: $P6_422$ (left) or $P6_222$ (right)
Unit cell	4.9124 Å x 4.9124 Å x 5.404 Å at 25° C Three molecules per unit
Physical properties Electric and light	Piezoelectric (strongly), Pyroelectric and Triboluminescent (emits light when struck) Birefringence: 0.009
Physical properties Cleavage and fracture	Does not break easily but can be cleaved in several directions. Conchoidal fracture
Physical properties	Hardness: 7 on Moh's scale Specific gravity: 2.65 Streak: white Brittle in crystals, tough in massive form Specific gravity 2.65 Melting point 1,410° C
Appearance	Luster: Vitreous (glassy) Some specimens are shiny or greasy Striations which sometimes appear on prism faces are perpendicular to the c axis.
Color	Colorless when pure and quartz is always almost pure. May be almost any color depending on impurities, especially cryptocrystalline varieties. Variations can occur within crystals.
Optical properties	Birefringence: 0.009 Dichroism: none Uniaxial positive Refractive indices $\omega = 1.544$ $\varepsilon = 1.55$
Transparency	Transparent to opaque

Some of the Persons who Helped with the Book

Doris and Harold Dibble at their Fiftieth Wedding Anniversary Party

Harold is a businessman with a lifelong interest in minerals. He graduated from the University of Michigan in mechanical engineering and then took postgraduate courses. He decided that the retail hardware offered more interesting challenges and opportunities than engineering and joined his father in the family business. He entered the Army in 1942 and was assigned to the Fire Control Instrument Repair Division of the Ordnance Department. He rose through the ranks to be in charge of instruction to both officers and enlisted men in that subject at The Ordnance School, Aberdeen Proving Ground, Maryland. In addition to instruction he was also responsible for the editing of various publications. In that capacity he was designated a Technical Information Officer.

In October 1942 he married Doris Isabel Gardner who had been a secretary to the scientists at Linde Air Products laboratory. Their marriage lasted over 58 years. They had three daughters: Virginia, Judith, and Sally. Doris joined him in the family business which, over the years, grew to two of the largest hardware stores in the area. Their life together was highly family oriented and their many different travels always had some aspect of mineral collecting. They attended the Detroit, Tucson, and Munich mineral shows as well as the Rochester Mineralogical Symposium. Their interest in art and the local artists was sufficient for the Buffalo Society of Artists to name them as trustees, in which capacity they have served over many years.

Sally Dibble

Harold and Doris retired from business in 1991 and established the Dibble Trust Fund Ltd., an educational fund for the science and the arts. They worked on several different projects designed to help interest others in the subject of minerals. This book arose from their dissatisfaction with the paucity of information about the variety and beauty of quartz crystals and the possible causes of the origin of all the different types.

Rudolph Muth is the IT person for the Dibble Trust Fund. For more than thirty four years he has been working with the leading edge of computer technology and organizational development. The team is grateful for his patience and sense of humor in making the book. He is also the owner manager of Adaptive Systems, a management and computer consulting service.

Rudolph Muth

Sally Dibble is a landscape designer and runs her business in the Boston, Massachusetts area. She had been an Arts major in college specializing in photography. Her previous work for the Trust Fund was in connection with the *Beauty in Minerals* video. In those photographs she brought out the diverse sculptural characteristics found in some mineral specimens.

Pamela Mills says she has thoroughly enjoyed working with the team at the Dibble Trust Fund. "It has been both an enlightening experience in learning about the beauty of quartz and a pleasure in contributing to the making of this book." She is responsible for the overall design of the book and the quality of its reproductions. She is the creative director of Mills & Associates.

Pamela Mills

Some of the Persons who Helped with the Book

Prof. Ichiro Sunagawa

Professor Ichiro Sunagawa graduated from the Department of Science, Tohoku University, Japan in 1947 and then joined the Geological Survey of Japan as a geologist moving up to a directorial position in 1970. He got his doctors degree in 1957 and became a professor of mineralogy at Tohoku University in 1971 where he stayed until retirement in 1988. He then became a Professor Emeritus.

During his long career he has written over 300 scientific papers in Japanese and over 200 in English plus 10 books in Japanese and 5 books in English. He has held many positions in and received many honors from various professional societies around the world. His main interests have been in crystallography, mineralogy and gemology with special emphasis on crystal growth.

Dr. Gerhard Niedermayr

Dr. Gerhard Niedermayr received his Doctorate from the University of Vienna in 1965. He has been the curator of the mineral collections since then and head of the Staatliches Edelsteininstitut (State Institute of Gemology) since 1980. He has done research in several areas with particular emphasis in alpine cleft type mineralizations. He has written about 350 scientific papers on various mineralogical subjects including gemology. Lectures at the Universities of Salzburg, Innesbruck and Graz are part of his schedule. A new copper-cadmium sulphate mineral,*Niedermayrite* has been named in his honor.

Prof. George Robinson

Professor George Robinson considers himself pretty much old-fashioned. "I just can't seem to understand why we spend so much time today learning about the natural sciences from CD-roms and the Internet, when the real world is just outside our "Windows". He is also curator at the A.E. Seaman Mineral Museum. He says, "being a curator for the museum reconnects me with the real world of minerals. It provides a basis and inspiration for teaching and research, and (please don't tell anyone) it is just plain fun!"

Jeffrey Scovil has been labeled *Photographer Extraordinaire* and that is a good summary of his work. His personality is a combination of several factors. He is very artistic, has a love of minerals, is indefatigable in his work schedules, an incessant traveler to see new specimens, meticulous in arrangement and lighting to bring out the best characteristics of the specimens and is interesting and pleasant to converse with.

Jeffrey Scovil

Dr. Ivan Bonev is Professor of Mineralogy at the Bulgarian Academy of Sciences in Sofia, Bulgaria. He is also representative to the International Mineralogical Association from Bulgaria. His main fields of interest are mineralogy, crystal growth and fluid inclusions. Other honored positions are: Senior residential Scientist and Vice President of the Bulgarian Mineralogical Society.

Richard Bideaux is well known in the mineral world. (Google turned up 512 references to him.) Much of his reputation is due to the meticulous accuracy of the books in the *Handbook of Mineralogy series* which describes all recognized mineral species. He has been General Partner in Mineral Data Publishing the company publishing the work. His co-authors in the series are John Anthony, Kenneth Bladh and Monte Nichols and they have worked for years on the books. The beautiful and very rare mineral Bideauxite has been named for him.

Bibliography

Blackburn, William H. and Dennen, William H. 1988,1994 *Principles of Mineralogy*
Wm C. Brown Publishers, Dubuque, IA, USA

Bloss, F. Donald 1994 *Crystallography and Crystal Chemistry*
Mineralogical Society of America Washington, DC, USA

Bonev, Ivan, 1983 *Review of the Bulgarian Geological Society, Vol XLIV, part 1*

Borofsky, Robert L., Whitmore, Robert and Chamberlain, Steven C. 2000 *Scepter Quartz Crystals from the Treasure Mountain Diamond Mine*
Rocks and Minerals July/Aug. 2000 pp 231-237

Cesbron, F., Lebrun, P. Le Cléac'h, J-M and Deville, J 2000
Quartz et Autres Minéraux de la Silice
Minéraux & Fossils Hors-série n 11

Chamberlain, Steven C. 1988 *On the Origin of Herkimer Diamonds*
Rocks and Minerals 1988 Nov/Dec pp 454-455

Chamberlain, Steven C., Robinson, George W. and Richards, Peter R. 1989
Mineralogy of the Alpine Veins near Sherbrooke, Quebec
Min. Record Vol. 20 pp 209-220

Chamberlain, Steven C. 1997 *Black-stemmed "Herkimer Diamond" Scepters*
Rocks and Minerals Mar/Apr p 21

Cohen, Alvin J. 1989 *New data on the Cause of Smoky and Amethystine color in Quartz*
Min. Rec. 1989 pp 365-367.

English, G. L. and Jensen, D.E. 1985 *Getting Acquainted with Minerals*
McGraw-Hill Book Co.

Frondel, C.1945 *History of the Quartz Oscillator-plate Industry 1941-1944*
Amer Min. vol. 30 pp 205-213

Frondel, C. 1962 *Dana'sThe System of Mineralogy Seventh Edition Vol. III Silica Minerals*
John Wiley and Sons, Inc.

Heaney, P. J. Prewitt C. T. & Gibbs, G. V. et al 1991 *Silica, Physical Behavior, Geochemistry and Materials Applications* Min. Soc. Of America

Jensen, David E. 1978 *Minerals of New York State*
W. F. Humphrey Press, Geneva, NY

Klein, Cornelis and Hurlbut, Cornelius S. Jr. 1993 *Manuel of Mineralogy*
John Wiley & Sons USA

Kostov, Ivan & Ruslan I.1999 *Crystal Habits of Minerals*
Academic Publishing House, Sofia, Bulgaria

Bibliography

Lasaga, Antinio C. 1998 *Kinetic Theory in the Earth Sciences*

Princeton University Press Princeton, NJ, USA
O'Keeffe, M. & Hyde, B. G. 1996 *Crystal Structures I, Patterns and Symmetry*
Mineralogical Society of America, Washington, DC, USA

Richards, R. Peter 1990 *The Origin of Faden Quartz*
Min. Rec. 1990 pp 191-201

Rossman, George 1991 Chapter 13 Heaney, P. J. Prewitt C. T. & Gibbs, G. V. et al 1991 *Silica, Physical Behavior, Geochemistry and Materials Applications* Min. Soc. Of America

Rykart, Rudolf second edition 1995 *Quarz - Monographie* [in German]
Ott Verlag Thun Switzerland

Sinkankas, John 1961 *Gemstones and Minerals*
D. Van Nostrand Co. Inc.

Sinkankas, John 1964 *Mineralogy*
Van Nostrand Reinhold Co., New York, NY

Sunagawa,I 1984 *Growth of Crystals in Nature* pp 63-105 in:
Materials Science of the Earth's Interior
Terra Scientific Pub. Co.

Sunagawa,I 1999 *Growth and Morphology in Crystals* pp 147-166
Mathematical Sciences #358 1993

Sunagawa,I 1987 *Morphology of Minerals* pp 509-587
Terra Scientific Pub. Co.

Sunagawa,I 1999 *Quartz Crystals Twinned After Brazil and Japan Laws* pp 23-36
Jour. of the Gemological Society of Japan 20 (1999).

Sunagawa,I 1987 *Surface Microtopography of Crystal Faces* pp 321-365 Morphology of Crystals
Terra Scientific Pub. Co.

Sunagawa,I 1992 *In Situ Investigation of Nucleation, Growth and Dissolution of Silicate Crystals at High Temperatures* pp 113-142 Annual Review Earth Planet Science 1992

Sunagawa,I 1982 *Morphology of Crystals in Relation to Growth Conditions* pp 127-134
Estusios geol. 38

Ulrich, William 1989 *The Quartz Crystals of Herkimer County*
Rocks and Minerals 1989 pp 108-122

Van Dyke, K. S. 1945 *The Piezoelectric Quartz Resonator*
Amer Min. vol. 30 pp 214-244

White, John S. 1990 *Amethyst Occurrences in the Eastern United States*
Min. Rec. 1990 pp 203-213

Index

A
Alpha Quartz .9
Altered Terminations .51
Amethyst .37
Ametrine .37
Artichoke Quartz .43

B
BCF .23
Becquerel, Henri .13
Berg, W.F. .24
Beta Quartz .9
Bilateral Parallel Quartz Druses .83
Blue Quartz .37
Bode, Rainer .4
Bonev, Ivan .19, 83, 95
Brazil Twinned Quartz .29, 31
Brewster Fringes .29
Brewster, David .29
Burton, K .23

C
Cabrera, N .23
Cathedral Quartz .49
Cipo Habit .45
Citrine .37
Cohen, Alvin J. .31
Conventional Crystal .21
Cristobalite .9
Crocodile Quartz .75
Crystal Face Surfaces .92
Crystal Growth .11, 19, 20
Crystal Shapes .39
Crystals with "Children" .48
Curie, Marie .13
Curie, Pierre .13

D
Dauphiné Twinned Quartz 25, 27, 29
Dibble, Doris .4
Druses .83

F
Faden Quartz .47
Fehlmann, M .68
Frank, C.F. .23
Friedlander Quartz .27

G
Galileo, Galilei .34
Gemstones .88
Globular Quartz .77
Goreva, Julia .37
Green Quartz .37
Growth Mechanisms .23
Gwindel Quartz .68

H
Handedness, left .15, 21
Handedness, right .15
Hartman, P. .20
Haüy, René .11
Haystack Quartz .76
Heaney, Peter .10
Hemimorphic type Crystals .83
Herkimer Diamonds .64
Hopper Quartz .56

I
Inclusions .72
Interrupted Crystals .62
Inverse Scepter Quartz .56
Iris Quartz .81

J
Jackstraw Quartz .76
Japan Law Habit .60

K
Keatite .10
Kossel, W .23

L
Lechatelierite .10
Lieber, Werner .29

M
Ma, Chi .37
Macromosaic .27
McCullough, David .66
Melanophlogite .10
Moganite .10
Moodie, A.F. .68
Müller-Krumbhaar .13
Mullis, J. .40
Muzo Habit .45

N
Niedermayr, Gerhard 4, 37, 54
Nucleation 19, 23, 29
Novelty Items
90

O
Odd Balls 80

P
Parallelogram 80
Péle de Jacaré 75
Perdok, W. G. 20
Periodic Bond Chain Mechanism 20
Phantom Crystals 74
Piezo-electric Effect 31
Pineapple Quartz 49
Pinetree Quartz 50
Plates of Tiered Crystals 81
Polymorphs .9
Post, Jeffrey 10
Properties of alpha Quartz 93

Q
Quartz "Diamonds" 64, 65

R
Rainbow Quartz 81
Richards, R.P. 47
Robinson, George 4, 95
Rocket Quartz Habit 43
Rome de l' Isle 11
Rossman, George 35,37
Rykart, Rudolf 27, 45, 47, 63, 68

S
Scepter Quartz 51
Scovil, Jeffrey 4, 95
Skeleton Quartz 54
Smoky Quartz 35
Steno, Nicolaus 11
Stishovite 9
Stransky, I.N. 23
Sunagawa, Ichiro 3-4, 19, 29, 60

T
Tabular Crystals 75
Tapering and Expanding Crystals . . .43
Tessin Habit 43
Tiered Crystals 80
Tridymite 9
Twinning 25
Types and Forms of Crystals 41

V
VKS 23
Volmer, M 23

W
White, John 77
Window Quartz 54

Y
Yasuda, T 60

Adsorbsion: The assimilation of gas, vapor or dissolved matter by the surface of a solid or liquid.

Angstrom Unit: Å is equivalent to one hundred millionth of a centimeter.

Anisotropic: Having different properties in different directions. This word is used in regards to crystal growth because crystals grow at various rates in different directions

Chiralty: Refers to the right or left handedness of a unit cell.

Epitaxial Lateral Overgrowth (ELO): This term, however, is not ideal. The word *epitaxial* indicates that the ***c*** axis of the new growth is the same or parallel to the ***c*** axis of the original crystal. Apparently the word *lateral* in this term originated wiith people who were developing chips for computers and other electronic devices and were very interested in lateral growth on flat surfaces. Actually there is equal growth in all directions possible, as the diagrams they use to illustrate this phenomenon clearly show. The word lateral should be omitted.

Epitaxy: The growth of the crystals of one mineral on the crystal face of another mineral, such that the unit cells of both minerals have the same orientation.

Euhedral: A crystal having faces which are well developed.

Isotropic: Identical in all directions.

Hydrothermal: Of or pertaining to solutions coming from hot magma.

Interstitial: Of, pertaining to or occurring in one of the small spaces between atoms.

Lamellae: Thin layers. Usually layers of sheet-like materials, often of alternating chemical or physical properties or substances.

Molecule: An electrically neutral group of atoms. It has all the electrons needed by the atoms of the group without having a surplus.

Morphology: As used by crystallographers is a word for overall shape

Nappe: Is a package of earth which is not in its place, which rests on a chance substratum, on a stratum which is not its original stratum.

Nucleation: The bringing together of particles which can act as a center of growth

Parameters: Variable or fixed boundaries or limits.

Polymorph: From the Greek *poly* meaning many and *morph* meaning shape or form. It refers to the same substance being assembled in different ways which results in different appearance and/or characteristics.

Polysynthetic Twinning: Means the multiple twinning of alternate layers. The word comes from *poly* (for many) and *synthetis*.(the combination of separate elements or substances to form a coherent whole) and not from the word synthetic (of artificial origin).

Re-entrant Angle:

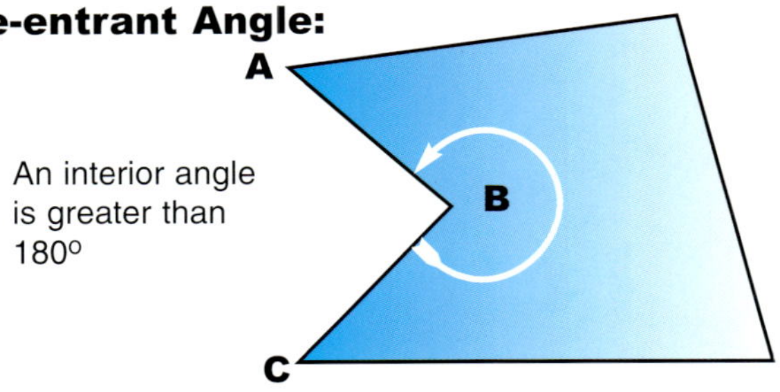

Saturated Solution: A solution that is holding as much of a substance as it can at that particular temperature and pressure. Super-saturation: a condition of a solution that has more of a substance in it than it can continue to hold at that temperature and pressure.

Triboluminescent: Short definition: Gives off sparks or light when sawed or struck. Long definition: Gives off light due to mechanical stress on the crystal lattice.

2DNG: Two dimensional nucleation and growth. This is a term used by crystallographers. Actually the growth referred to is thin, sheet-like, but it does have a third dimension even if it is only a few angstroms thick. They are not talking about nucleation either because only an attached unit cell is needed for that. The term really refers to leaf-like growth.